Religion and Technology

Man and Religion Series
Part 7
Christianity and Our Culture

The Religious Education Press
a member of the Pergamon Group
Headington Hill Hall
Oxford

Religion and Technology

Keith Wilkes

Keith Wilkes
is
Diocesan Director
of Religious Education
and Lecturer in Theology
College of St Matthias
Bristol

Design and Cover
Keith Clements

Man and Religion Series
General Editor Ronald Dingwall
Executive Editor Gordon Hawes
Art Director Keith Clements

The Religious Education Press
a member of the Pergamon Group
Oxford New York Toronto
Sydney Braunschweig

First published 1972

Made and printed in Great Britain by
A Wheaton and Company
Exeter

Library of Congress Catalog Card No. 79-134732

08 006804 9 (flexicover)
08 016560 5 (hard cover)

Contents

Foreword

The impact of our rapidly developing sciences on the age-old spiritual and moral problems facing us as human beings is an enormous and complex subject. Yet it is one that we must all face if we are to find our way through the technological wilderness growing up around us, which could, if left to itself, strangle man's essential spiritual qualities.

This book sets out clearly and freshly the main areas where technology encroaches on our community, fields of choice, responsibility and enjoyment. On many questions of personal judgment, each reader will want to make up his own mind, and there are bound to be views and opinions expressed in this book on which some will disagree. But this is the main quality and value of the Rev. Keith Wilkes' approach. For to pose the questions is always the first task.

Everyone who reads this book will be encouraged to examine afresh the objectives towards which we should be directing ourselves, and the values we must uphold in doing so.

The Rt Hon. Anthony Wedgwood Benn, P.C., M P.

Introduction

This book is a companion volume to **Religion and the Sciences**, published in the same series. It aims to give some indication of the nature of technology and the vital part which man's tools and techniques have played and will increasingly play in his social and cultural development.

In Part One, an attempt is made to show how man's cultural life and world-view have been modified by technical changes, and how religious beliefs have played an important part in those changes—particularly in the emergence of the city and the industrial revolution. At the same time, questions are raised about technology and change which are of great importance today, and certain key issues—the idea of progress, the nature of culture, the philosophy of technique and the human implications of modern technology—are emphasised.

In Part Two, a limited selection of modern questions is outlined in the light of the earlier discussion. The nature and form of these questions are designed to show clearly the bearing of religious and moral convictions upon them, so that the debate about the complex relation between religion and technology today may be appreciated through problems with which all of us should be familiar. In the final chapter, a sketch is given of the main areas in which modern technology impinges on Christian faith and life.

Any adequate discussion of the history and nature of technology, let alone of social and cultural change, requires careful and extensive treatment of a kind which would vitiate the purpose of this book. At the risk of oversimplification and generalisation, the slighter treatment attempted here may be justified by the importance to be attached to these topics being encountered more widely at school, college and elsewhere. The aim here has been to provide no more than an outline introduction, and simply to identify the range of issues at stake. The approach is intended to encourage readers to make their own investigation, and to formulate responsible attitudes towards the problems they will meet.

There can be no more important field of investigation for any satisfactory syllabus of religious or liberal education today than the interaction of religion and technology. There are certainly few fields in which the ordinary Christian needs more understanding and guidance. If this book helps the reader to discover a fresh perspective on the problems confronting a fast-changing society, its purpose will have been fulfilled.

I must record my gratitude to the former Minister of Technology, the Rt Hon. Anthony Wedgwood Benn, for his kindness in writing the Foreword for this book, and for his encouragement to my raising issues with which he is particularly concerned. I should like to thank those who have given advice and constructive criticism during the preparation of this work, especially Ronald Dingwall (General Editor, Man and Religion Series), the Reverend David Stacey (Wesley College, Bristol), the Reverend Donald Whittle (Newton Park College, Bath), Mrs Sybil Hodge (School of Education, University of Bristol), and Armine Wright (College of St Matthias, Bristol). I have adopted many of their suggestions, but I am alone responsible for the remaining imperfections. I should like to acknowledge the contribution made to the ideas in this book by students of the Universities of Birmingham and Aston and of the College of St Matthias. Some of the material has been used in extra-mural classes sponsored by the University of Bristol.

My thanks are due in particular to my wife for many of the facts and arguments in the book, and for her endless patience and encouragement. This book is dedicated to John and Susan, for whom it was really written, in the hope that they will be able to live with dignity, enjoyment and care in the world of tomorrow.

Keith Wilkes
Bristol
April 1970

Acknowledgments

Thanks are due to the following publishers for permission to reproduce copyright material:

Routledge and Kegan Paul for **Culture, Industrialisation and Education** by G. H. Bantock; *CA Watts and Co.* for **Progress and Archaeology** by V. G. Childe; *Cambridge University Press* for **The Wheelwright's Shop** by George Sturt; *Penguin Books* and *Scientific American Inc* for **Technology and Economic Development** article by Kingsley Davis; also for illustration from article **Population,** Kingsley Davis © 1963 by *Scientific American Inc.; Oxford University Press* for **A Sand County Almanac** by Aldo Leopold; *Cobbett Press* for **Men, Machines and History** by S. L. Lilley; *S.C.M. Press* for **Letters and Papers from Prison** by Dietrich Bonhoeffer; *George Allen and Unwin* for **Human Documents of the Industrial Revolution**, edited by E. Royston Pike (extract from Peter Gaskell, **The Labouring Population of England**); the *World Council of Churches* for **The Uppsala Report**; *Division of Christian Education of the National Council of the Churches of Christ of the United States of America* for quotations from the **Revised Standard Version of the Bible**.

Part One
TECHNOLOGY, SOCIETY AND RELIGION

A
What is Technology?

I
Tool-using and tool-making

On his famous voyage in the **Beagle** to the Galapagos Islands in 1835, Charles Darwin noticed a finch using the spines of cactus plants or twigs of wood held lengthwise in its beak to poke out insects embedded in the branches of trunks of trees. He called it *Cactospiza*. It has since been noted that the bird selects twigs of suitable length and stiffness for the task and rejects others—a clear case of intelligent tool-utilisation.

Nearer home, the Greater Spotted Woodpecker is known to construct a vice in a tree-trunk by pecking with its beak. It then inserts a pine cone to hold it firmly while removing the seeds.

Among the mammals, the activities of the Canadian beaver are well-known. In addition to felling logs to construct dams, the beaver scrapes canals alongside its pools down which logs are pushed to be stockpiled for the bark during winter. The Southern Sea otter collects molluscs from the sea bed, breaking each one open by holding the shell in its forepaws and striking it on a stone brought to the surface with the mollusc and balanced on its stomach. The fascination of monkeys and chimpanzees in captivity rests in part on their ability to use tools, even eating utensils.

Numerous experiments have been performed with apes to test their skill at tool-using and the way in which they learn various techniques. These experiments have established that skill is akin to technique, that it can be learned in adolescence, and that its development is bound up with a number of interrelated biological and psychological factors. Behavioural studies indicate that at least four factors are involved: (a) powers of sensory perception; (b) the ability to coordinate sensory perception; (c) physical capacity; and (d) the demands of the environment. The use of different kinds of tools is seen by biologists as one means by which an individual organism adapts to or alters its environment.

These behavioural investigations cannot answer the question why the *hominidae*—man and his immediate predecessors—became toolmakers. No complete explanation has been offered for the origin of this capacity. At best it may be said that man's urge to make and use tools depends upon his psychological make-up and his sophisticated mental apparatus. These little explored elements of human nature hold the key to man's urge to make, organise, operate and control.

In the era of pre-history—before written records are available—the precursors of man probably only used tools occasionally for cutting (axes), hunting (spears and hatchets), painting or making fire. Their normal life seems to have been in the caves and scrub-lands rather more than in the woods. The systematic manufacture of tools such as flint axe-heads could only begin at the stage of his biological evolution when the cortex of the brain became sufficiently complex to make it possible. Then factors such as the need for and availability of flint would make tool manufacture a habit. Tool-making appears to have begun among the predecessors of man at similar times in widely separate parts of the world such as Northern Europe, South Africa, Java and China.

The decisive distinction between the tool-making of man and the tool-utilisation of other creatures lies in the fact that the *hominidae* were able to imagine the end-product before the tool was made; something more than trial and error was involved in his efforts. Man the maker—*homo faber*—is thus something of an artist as Aristotle defined him; he has the 'capacity to conceptualise the result to be produced before its realisation in the material'. The importance of imagination in technology is therefore established at the outset: its continuing importance throughout the history of technology and in every aspect of it must be grasped before the true nature of technology can be understood. While it is well mixed with the empirical procedures of experiment—trial and error and practical modification—the use of the imagination is always present. It is to be seen in the axe and the bone needle, in the invention and development of the ship and the wheel, in the marvellous devices of Leonardo da Vinci as well as in the Concorde and Severn Bridge. It is by the particular use of his imaginative powers that the human organism adapts to many kinds of environment such as the extremes of cold encountered at the Polar regions, the world beneath the ocean or the hostile situation on other planets and in outer space. It is the human inspiration which feeds man's desire to explore—as the American Moon programme vividly shows. By the use of his imagination and skill man can now exercise increasing control over his normal environment on earth; making rain, using water for large-scale irrigation schemes, providing a new urban landscape or using new media of communication. The use of the laser beam as a precision tool or of the infra-red camera in the earth satellite may appear at first sight to be far removed from the hand axes or pots of primitive man, but all are tools, and all have derived from the imagination and skill of man the maker.

2

Magic

The French writer Jacques Ellul[1] has distinguished two paths of human technical development. The technique of man the maker who invents and uses tools, builds houses, bridges, machines, organises his society and extends his natural capacities in various ways, and, alongside this, the technique of the spiritual order, especially magic.

Like material technology, magic is a conscious attempt to control the environment in which man lives. By performing a magical act—saying a prayer, sacrificing a chicken, or reciting an incantation in prescribed form—it is believed that a supernatural power must respond in an expected way, usually by intervening in some natural process as desired by the operator In such a closed system, magic is obviously a technique like the introduction of an irrigation wheel or the use of a medicine. Magic offers a way to placate hostile spirits, to protect man and his society—living or departed—from unknown or inexplicable disaster. But magic is also the form of expression of a sense of life, that is, an attempt to come to terms with the mysteries surrounding life and death, and therefore comparable to science. Magical forms—and forms in several types of religion—are fixed elements in a community's understanding and life. They 'compel' a response from the spiritual powers which surround man in accordance with the mystery known to the magician, witch-doctor or holy man.

Magical techniques in any given community have not followed the same development as material techniques, instead they have remained largely unaltered. There is plenty of evidence to indicate that magical and religious beliefs have often interfered with the advance of material technology. Hostile attitudes towards artificial methods of birth control have been expressed in

[1]J. Ellul **Technological Society** Cape 1963

parts of India and Latin America on the grounds that time-honoured magical procedures to achieve the same ends are already in use, or that such interference is contrary to the will of the appropriate spirit. The hostility of the government and ruling party in China to all forms of religion is expressed partly because the religions of China have generally opposed the adoption of simple engineering or agricultural techniques needed to accelerate production and improve the lot of the workers.

Material technology has become a progression of discoveries, inventions and skills built upon one another. Magical techniques offer only one constant system, in any given community, and only new beginnings in a new civilisation. But in point of fact they appear in every society, including the most technically advanced, with their unquestioned myths, rituals, rules and panaceas. It is perhaps inevitable that religion has become confused with magic again and again in every society, or that magical and material technologies have been given a guise as authentic religion. The people of Israel in the time of the prophet Jeremiah (sixth century B.C.) had turned their temple—a symbol of past assurance—into a magical object which would, they believed, prevent their being overrun and exiled. Christian sacraments have often been regarded in magical terms, so have prayers—especially prayers of intercession and supplication. Congregations are still urged to efforts of devotion and petition on the grounds that God will be persuaded to intervene in the affairs of men if the right prayers are said with enough intensity and persistence. Such a view of prayer presents a totally inadequate understanding of the Christian or Jewish view of man's relation to God, and is a widespread modern example of magic disguised as religion. The forms of magic in modern industrial society reappear in the newspaper horoscope, and even more pervasively in the advertisers' recipes for success, protection and instant happiness. Who does not bow down before the god of super whiteness or the spirits of high performance petrol? Who is not sometimes persuaded that happiness and control of the future is assured by sexual charm, slimming aids or life insurance?

Modern faith in technology comes close at times to a naïve belief in the unknown, with all the obedience of a religion and all the operational emphasis of magic—if you get it right it will work. Such is the nature of man that he is torn between reason and faith, and between technology and magic at his most reasonable and his most religious moments. For all the technological progress that he has made in the industrial world, our society appears to be under just as great pressure to revert to magical attitudes and procedures as any before us, even if the magic has a different form. Technological man is not a whit less superstitious—or potentially religious—than his Miocene ancestors.

3
Craft and Art

The evidence available to us of early technology—and medieval and even nineteenth-century technology in certain fields—comes from archaeological finds and, later, pictures or sculpture rather than written records. In the earliest periods writing had not been invented, but even as recently as the Middle Ages it was used for reflection and administrative use rather than for technical description. The methods of archaeological investigation have to be used to plot man's technical history. The fact that many practical men among the ancients and medievals were not writers meant that much of their knowledge was passed on by example and word of mouth and not committed to tablet or papyrus. In this way the various technologies established the idea of apprenticeship and craft tradition. Learning took place by practical training on the job and by constant experience. The traditions of craft work still bear the marks of this method of transmission, and the ideas of secrecy, craft-signs drawn from ancient languages and known only to initiates, and close-knit guilds still persist in some quarters.

The craft practices of the ancient world may be grouped about both natural and non-natural needs. Natural needs include farming, irrigation, cloth-making, pottery, cooking; non-natural needs include perfumery, glass manufacture, the fine arts. Frequently these non-natural crafts were used to imitate nature—as in dyeing cloth and manufacturing make-up, beads and trinkets—and to glorify the deity in places of worship. Natural crafts were also closely associated with religious practices. It was necessary to employ the techniques of farming in relation to the cycle of nature—and it was firmly believed that this was a matter under the control of the gods. Thus seedtime and harvest were linked to ritual acts and special days chosen by the priests. Similarly the practice of medicine and the beginning and completion of buildings were associated with prayers and offerings. Even the lighting of furnaces was performed on certain days with special ritual acts.

In a world-view which accepted a very close relationship between events on earth and in the heavens—sometimes in the stars themselves—such practices were both logical and necessary. Here lie the origin of many modern religious festivals, suitably Christianised by the Church over the centuries; Easter, Rogationtide, Harvest, Christmas. In this world-view is to be found the basis of alchemy and astrology, and the explanation for the length of time it took for the craftsmen to begin to reflect upon the material explanation of their trades. Such reflection scarcely began before the sixth century B.C., and was eclipsed for many centuries in the Dark Ages. It was not until A.D. 1850 that scientific technology emerged—that is to say that the systematic understanding of processes began to be deliberately applied as techniques. The craft tradition and its pattern of apprenticeship learning is still a major factor in the reluctance of certain sectors of industry to take notice of scientific research and modern technical advances. There are still occasional reactions—especially in the churches—to the so-called loss of craftsmanship in the twentieth century. There will always be a tendency for conservative and religious men to hanker after the simple, homogeneous world-view of the past. But craftsmanship, like religion, has to come to terms with modern science and technology. (For a more detailed discussion of Alchemy and of the rise of modern science, see **Religion and the Sciences** Part I.)

The range of activities which have already been indicated as falling under the overall heading of technology is by no means exhaustive. In particular, little reference has been made to the place of technology in the arts; sculpture, painting, drama, dance, music and the like. Any satisfactory discussion of human technology must refer to this area as well as to the manufacture and use of implements, machines or social systems.

Sir Kenneth Clark[1] has convincingly shown that in the era before the development of popular printing and widespread literacy, the artist and the sculptor held a firm monopoly in the field of popular communication. Although the emergence of the art-form of drama from the liturgy of the church had occurred in the Middle Ages with the miracle and porch plays, and the wandering Franciscan friars and Dominican monks had some success with the public sermon in the thirteenth and fourteenth centuries, the visual artist and the architect clearly exercised most influence up to the fifteenth century as communicators of ideas, beliefs and values. With the invention of printing and its later commercial exploitation in Europe, vernacular texts and realistic woodcut illustrations became available. Communication then became concentrated increasingly in words. From the sixteenth century, the greatest enrichment to European civilisation was made in literature rather than in painting, sculpture or architecture. In spite of the achievements of the Dutch school of painters with its Rubens and Rembrandt, there is nothing to compare with the literary art of Erasmus of Rotterdam, Montaigne or Shakespeare. It is only in recent years that sculpture has been revived and used in new ways to communicate a fresh

[1]K. Clark **Civilisation** *John Murray* 1969

vision of reality. Only in this century has architecture departed from the imitation of great styles of the past. These developments show a direct dependence on the use of new tools and techniques, and the availability of new materials, as well as on a new understanding of man and society.

Lewis Mumford, the architectural and social critic, has suggested that while art and technology are rooted in man's use of his own body, they can be distinguished. Art, he maintains, is primarily the domain of the person; its function is to widen the province of the personality in such a way that feelings, attitudes, values and emotions can be transmitted between people. Thus although the different arts require different techniques for their accomplishment—of paint manufacture and mixing, brushwork, mounting in the case of painting, to give one example—they belong nonetheless to a common level of human activity which is different from that concerned with providing for man's material survival.

From the earliest periods of human life art is associated with religion. No distinction was made in prehistoric times between the sacred and secular areas of life that are readily marked out today. Primitive worship, prayer, ritual dancing, singing were full of art—and of special techniques too. The cave paintings at Lascaux in Southern France show bison and horses which are believed to have been symbols in the practice of the religious rites of the Aurignacian people. Animal masks also found in the caves were probably worn in ritual dances. The decoration of weapons used in the chase was made to bring the blessing of tribal deities upon the hunt. The decoration of aircraft in wartime and the devotions of fighting men express the same primitive religious motive today. The wearing of charms, beads, anklets or rings—often made of very costly and rare materials—appears to be a timeless attempt to ward off danger in a hostile world, whether this is the African amulet of 5000 B.C. or the crucifix hanging round a twentieth-century teenager's neck.

The legend (or ancestral myth or saga) and the dance take their origin in the religious life of every people from the Viking invaders of Northern Europe to the Polynesians of the Gilbert and Ellice Islands in the South Pacific.[1] Stories and dances have provided effective ways of uniting groups of people in tribal, racial or national structures. They explain the origins of the group and give it an identity in the vastness of the world. They often attempt to provide a coherent account of the mysteries of life in a universe penetrated by the spiritual powers of good and evil. No more vivid sample of such stories can be found than those recorded in the first eleven chapters of the book of Genesis. Such stories, when retold by word of mouth or in the written form, involved important techniques in the telling, and performed a technical function in the group, nation and religious gathering. At the same time they may properly be described as significant forms of art with a universal appeal.

The greatest single technique common to all art forms is that of making and using symbols. These may be visual or audible, touch-symbols or smell-symbols. They require actor and audience, painter and viewer, dancer and onlooker, speaker and hearer, agent and recipient. The making of symbols, possibly even more than the capacity to make tools, belongs essentially to **human** existence. It binds a man to his neighbours in a social and cultural complex, it unites the past with the present, and enables man to think of the future. It discloses the hidden meaning of life and its untouched possibilities.

In essence, man's life of symbols—in contrast to his life as a technician—relates to internal changes rather than external ones. Symbols express man's awareness of himself and his significance; they express his inner response to the outside world, even though they may be mathematical or scientific in character. The symbol can personalise the world in a way which material technology can rarely do. Technology often creates an impersonal situation such as a self-correcting computer, or a relentless production line. Man's role in relation to much technology is that of a driver or operator, or at the worst a displaced person

[1]See Arthur Grimble **A Pattern of Islands** *Murray* 1952

in a vast, soulless system; but not in relation to symbols. One of the greatest challenges to humanity at the present time is to relate positively man's symbolic life to the world of the machine and the organisation. Industrial man has made little progress so far in relating the world of art and religion to the world of work and economics. But, at best, technology and art can combine in massive statements of faith such as the pyramids of Egypt, the new Opera house in Sydney or the Metropolitan cathedral in Liverpool. Sometimes they achieve the glory to be found in the cathedrals at Chartres and Santa Sophia in Constantinople—or the mysterious beauty of the **Book of Kells** or Henry Moore's **Madonna**. These possibilities have to be set against those awful pictures of twisted trees in a dustbowl, of vultures over the battlefield, or of shoddy shop windows in endless tenement streets.

Although art and technology may, as Mumford suggests, be distinguished, they are inextricably interwoven. The making and use of effective symbols is a technical affair—in speech, writing, number, painting, or design. The growing problem of the past two hundred years has been to achieve the proper balance in this interrelationship: to discover how technology may best serve man's symbolic, artistic, moral and religious life. One great issue for each member of industrial society is to discover how to live a full, inner personal life of integrity and creativity in the massive world of technology. The problem has often been explored and stated: by implication as much in Rembrandt's self-portraits and Van Gogh's fields and flowers as in Picasso's **Guernica**. It is variously expressed in the novels of Charles Dickens, D. H. Lawrence and Samuel Butler; in the poems of William Blake, Matthew Arnold and T. S. Eliot, and it is faced directly in Marlowe's **Faustus** and the Theatre of the Absurd. It is to be found at the heart of the Christian fatih—when the instruments for killing the worker from Nazareth became the symbol of hope and new life for those who recognised his risen presence. In the last resort it does not appear that the life of man the tool-maker and man the artist can be finally separated, since man—like the

God in whom Christians believe—is a creative being. (See here **Art and Life**, by Keith Clements, **Man and Religion Series** Part 7.)

4
Types of Technology

To begin with, technology had to do with the basic things of existence—gathering food, providing protection from enemies or the elements, establishing simple social life, and finding a meaningful place in a mysterious universe. Thus man invented axes, spears, knives and picks; he used materials such as stone, wood and bone; he learnt to cure skins and make rope, and to build in wood, wattle and mud. Simple cookery developed as man controlled the use of fire with spit, and discovered the taste of cooked roots and herbs. The mysteries of life evoked a response in painting, carving, the making of masks and special ritual garments, and the wearing of charms.

Life in later communities saw the techniques of the herdsman and the farmer developed, with pottery, textiles, mats, baskets and buildings of stone and brick. Agriculture stimulated the irrigation of fertile valleys in sub-tropical regions like the Nile delta. Here are the origins of civil engineering and textile technology.

The discovery of metal and the means of smelting ores to produce it in quantity led to the technologies of mining, quarrying, metal refining and the manufacture of alloys, and of casting tools and weapons. The decoration of metal, ivory and wood could then proceed; thus a whole new range of arts and crafts became possible. With new metals grew new ways of warfare, better manufacturing tools and methods of transport.

Human technique embraces the varied modes of transport and communication from the ox or camel lurching over ancient tracks, to Roman legions marching over slave-built roads and bridges. It covers the invention of the wheel and wagon, the ship, the stage-coach, the automobile and the jet airbus. Each section of transport engineering—as well as the planning of transport systems—is now a large field of technological activity in its own right. The diesel locomotive and the hovercraft are tools, just as were the horse and cart before them: as such they require technical skill to operate and maintain. Travelling man developed the arts and tools of navigation: the astrolabe, compass, map, chronometer, radio and radar.

The civilisations of the ancient near East fostered the birth of writing and number. From early symbols for weights and measures mathematics and astronomy emerged. Now we have many scientific languages and symbols for weight, measure, energy and time; we have the techniques of accounting and computing, which multiply man's capacity to calculate beyond belief.

Technology deals with fuels and sources of power: wood, coal, gas, oil, nuclear fuels and water. It deals with the machines and devices for converting one form of energy into another, such as water-power into grinding motion or nuclear power into electricity. It includes the various forms of engine and motor which can be used to concentrate energy into a useful form as it is required, whether this is by a Roman galley-slave or a Rolls-Royce turbofan jet engine.

Technology covers the whole of the chemical industry from food preservation and preparation (of cavemen or top chefs), the manufacture of perfumes, dyes and paints, to detergents, fertilisers, plastics and pharmaceuticals.

Nowadays new technologies are constantly appearing: urban and regional planning, computer technology, management techniques and advertising—new forms of social engineering and administration for politicians and business men to use. Alongside these are to be found the new medical technologies of transplant surgery, the use of artificial organs, chemotherapy and psychotherapy—a far cry from the medicaments of the ancient world and even from the crudities of nineteenth-century surgery or physic.

Technology—its tools, systems, machines, structures and devices—therefore touches every human concern. It is a basic activity of man. This should be more evident to us today than to our forefathers because of the extent of technical activity, its complexity, and the variety of technologies available to us which influence our way of life.

It is possible to group technologies in various ways. A distinction between 'natural' and 'non-natural' types has already been suggested above. Another pattern of classification is that adopted in the five volume **Oxford History of Technology**: in relation to *Primary Production* (food, fuels, raw materials); *Manufacture* (textiles, chemicals, machines and tools); *Transport and Communication* (roads, railways, vehicles, ships); *Fine Arts*; *Human and Social Techniques*.

Another approach has been to group technological operations together under these headings: (a) those aimed at raising the material standard of living; (b) those aimed at increasing leisure, or eliminating dull and unnecessary work; (c) those aimed at replacing less efficient technology by more efficient. This scheme has the advantage of bringing out clearly certain features of technology common to all ages.

First, it indicates the universal character of technological activity. As with scientific activity, with which modern technology is increasingly involved, technology is a world wide phenomenon—a human phenomenon. It has the effect of bringing men together into cooperation more and more. Although technology has often been used divisively in war, economic competition or with regard to class and social distinctions, it is now an even stronger uniting influence. Under its influence, the world has become a small planet, it takes a few hours only to fly round it. Telecommunications have put the Antipodes on the end of a telephone. The African labourer and the English housewife, the Indian weaver and the Hong Kong business man share with the German industrial chemist and the American farmer a common interdependence and increasingly a common set of values and way of life.

Secondly, this outline classification shows the relation of all technology to human life. Work and leisure are undoubted 'goods', but they have to be properly directed and used. Technology may enter many kinds of leisure; driving a car, playing golf, watching television or going to a dance—but the satisfactions of leisure are more than the sum of the technology involved. They are emotional, aesthetic, moral and personal as well as technical.

Thirdly, this analysis of human techniques shows how the drive for better technology develops. Technology can give man better health, better protection, more adaptability in his environment. It increases his range of choice and freedom of action. It can widen the satisfactions of work and leisure open to him, and thus generally enhance his position in the universe. It is the advance of technology, especially in recent years, which has enabled man to 'come of age'. In the application of technology man becomes more dependent upon his fellows all over the world. He performs specialist tasks from which others (whom he does not know and will never see) benefit. At the same time he gains greater independence and responsibility, acquiring a new stature compared with his forefathers farming their strips of land. Through his technology, man can become far more godlike—but he can also become more demoniacal. He can master his own existence within certain biological limits—he can cooperate effectively with others to transform his planet and explore his universe in all directions and at every level from the microscopic to the astronomic—and he can far more readily destroy himself and countless others. The contrast between the atomic reactor and the atomic bomb—or between the exploration of Mars and the Vietnam war—makes this possibility absolutely clear. John F. Kennedy said:

> *Never before has man had such a capacity to control his own environment, to end thirst and hunger, to conquer poverty and disease, to banish illiteracy and massive human misery. We have the power to make this the best generation of mankind, or to make it the last.*

However we classify the different technologies, it remains a fact that few people fully understand their significance in the present situation. The increasing effect of technology, in the Afro-Asian countries as well as in Europe, China and the Americas, demands that we make the effort to understand and master it, or run the risk of being carried away by our own achievement.

5
Technology as Applied Science

The invention of the thermionic valve by J. A. Fleming in 1904 lies behind the whole of modern electronics. Making use of the 'Edison effect'—the emission of a stream of electrons from a heated filament in a vacuum—Fleming and Lee de Forest produced various valves for the rectification and amplification of alternating currents. These devices became the basis of wireless communication. The recent development of the solid-state transistor, which has largely replaced the valve in electronic circuitry, has made possible many refinements and the miniaturisation of electronic apparatus. Radar is but one of many fields in which the science of electronics has been systematically applied with considerable social consequences.

This example introduces a feature of modern technology which has not so far been mentioned—the deliberate application of scientific knowledge to practical use. Early technology had little scientific foundation. Even the internal combustion engine and the first aeroplanes owed more to the empirical, trial-and-error approach than to scientific theory. There were no general theories or specific concepts associated with the solution of practical problems in most areas of technology for thousands of years. The beginnings of such concepts and intellectual map-work have their origin in the cities of Mesopotamia, Egypt and later Greece, employing the skills of writing and sign-making. Following the invention of graphic symbols and the adoption of mathematical standards of weighing and measuring, general and abstract reflection became possible and had its first applications in city building and simple astronomy. Babylonian astronomy had a profound influence on Greek science, thanks to the conquests and interests of Alexander the Great (355–323 B.C.) In the Babylonian schemes, specific empirical data is limited, but it was publicly recorded and provided a basis for theoretical predictions which led to practical consequences such as the production of a calendar.

From 550 B.C. onwards, the Babylonian mathematics had been assimilated by the Greeks, who supplemented it with geometry and—after Aristotle and his school—a growing interest in scientific description and classification. It may fairly be said that the Greeks were the first in the western world to develop a coherent scientific activity in their society, as the Chinese had done in the fifth century B.C. Yet astonishing as it may seem, Greek science did not bring many practical advances in technology. This may be attributed to the influence of religious and philosophical ideas which were preoccupied with the life of the soul or mind and the pursuit of political reason. Further, the mechanical arts, as Xenophon noted in **Oeconomicus**, carried a social stigma, and were supposed to lead to physical degeneration and 'deterioration of the soul'! This attitude—exemplified in the writings of Plato (427–347 B.B.)—effectively prevented the inventions of Archimedes and Hero from going far. Devices like the steam turbine and the gear might have transformed industry and agriculture. The abundance of cheap power units in the form of slaves and the consequent lack of commercial incentive also kept such schemes in the inventor's backyard. The technical devices which the Greeks did develop—incendiary arrows, fire-ships, elementary explosives on the one hand, the lighthouses and water-pumps on the other—were stimulated by warfare and social demand respectively. These two factors have always been the most potent in stimulating technical advance and application. For the rest, Greek science corresponded to wisdom as a self-justifying way of life, and not as a means of human control of the external world.

In ancient Rome technology advanced along certain lines determined by the national interest in order, warfare and good living, but science as such was largely ignored. Early science in Babylon and Greece was an activity of a small number of men with leisure and interest, who had the education and resources to observe, speculate and record their findings. The extensive scientific writings of the Greeks were preserved in Greek and Arabic translation to be rediscovered by later generations. Nevertheless science remained a socially peripheral activity before the eighteenth century, with little practical application.

Much—though by no means all—modern technology is scientifically based; that is to say it is the systematic application of scientific knowledge to a practical end. Whereas for centuries in the steel-making industry the composition of various recipes were essentially craft-skilis learnt over many years by trial and error and handed down from generation to generation, nowadays the modern steelmaker relies on highly trained metallurgists with expert knowledge of the properties of materials and methods of creating them. For this reason, the iron and steel industry, like other major manufacturing concerns, gives substantial support to fundamental scientific research. This kind of development is altering the pace of technical change on all sides. It is also creating a shortage of trained scientific manpower in modern industry. This shortage, together with lack of capital resources to put new processes into production, seem to be the present limiting factors to the pace of technological advance in western countries. Where these factors are overcome—as in the U.S. or U.S.S.R. space programmes or in the Anglo-French *Concorde* project—enormous strides have been made with many side-benefits affecting the

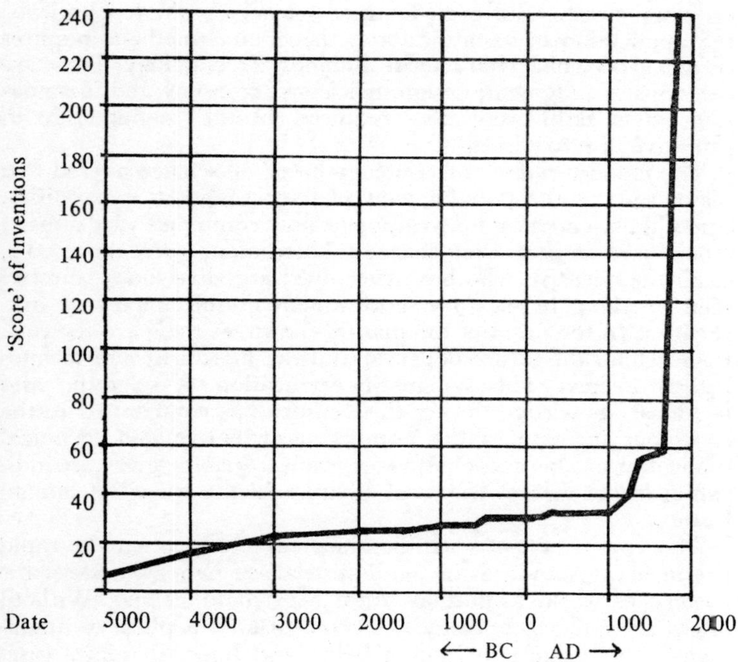

figure 1

The graph indicates the number of fundamental types of machines and tools at man's disposal at any given date. (After S. L. Lilley Men, Machines and History *Cobbett Press* 1948.)

ordinary citizen. But even in these science-dominated projects, the application of scientific data, theory and methods requires certain skills which characterise all kinds of technology. These are the skills of judgment, design, efficiency, economy and imagination: such skills were also required in the manufacture of primitive flint axe-heads.

The modern phase of technology based on science has led to a major shift in the popular view of science and its capabilities. Scientific research and development now command vast sums of money and an increasing share of skilled manpower. Some 90% of all the scientists who have ever lived are alive today, most of them working in industry, government establishments or universities. In the mind of the man-in-the-street these are the men upon whom the future depends, and the high priests of technological society. The systematic application of scientific and technical knowledge during this century has contributed to the enormous increase in the number of inventive and technical innovations. The accompanying graph (*figure 1*) gives an indication of the rate of technical advance in the period of human history.

The application of scientific knowledge has caused the rapid rise in living standards in the industrialised nations, and set the pattern of world aspirations from Mexico to Peking. Without doubt scientific technology has come to hold a place of fundamental importance in human belief and hope, in many cases displacing traditional religious convictions and longings. At all events, modern technology may fairly be described as the extension of scientific man, who is in many ways a different creature from his pre-scientific ancestors. This is a fact which Christians and others have yet to face.

6
The extension of Man

The biological evolution of any organism in response to changes in its environment is a long, slow business. In the case of organisms having an extensive life-cycle it takes many thousands of years. The invention of the simplest tools enabled man immediately to become the most adaptable creature of all. Through his tool-making and elementary skills he could make fire and clothing which enabled him to live in hard, cold climates—even at the edge of the ice cap during the great Ice Age in the Northern Hemisphere up to between 11,000 and 9,000 B.C. Fire not only provided warmth, but a means of cooking otherwise unpalatable meats and roots; it gave light and protection in the dark against various enemies; it made possible the production of dyes and pigments for camouflage or decoration. With even the simplest flint or bone instruments, man was able to cut trees and saplings and kill game at a distance, he could then make clothes and construct shelters proof against the weather. In this way early man was able to spread into every climatic zone where he could find food.

With the development of language from various cries, grunts and gestures, man could communicate his skills to his neighbours and his young. By passing on these hard-won aptitudes and techniques, he was able to establish traditions, characteristic ways of life, social patterns and, eventually, cultures. The observation of isolated tribes in remote parts of the world today, such as the Amerindians of the Amazon or the Aborigines of Australia, has given a reasonable indication of the early stages of human technical and social development even in extremes of climatic conditions.

In any appraisal of man's technological development, it soon becomes apparent that basic human technology has enormously extended man's powers and adaptability. It has enabled one of the weakest of the animal species to become the master of both animate and inanimate creation, and given him the strongest position in nature with great powers for good or ill in relation to himself, his own species and his environment. The combination of various tools with human desires, skills, imagination and determination has placed man in an unique position. This is confirmed by a more detailed study of the technical development of any sector of the human race, such as, for example, that contained in the massive seven-volume **Science and Civilization in China** by Dr Joseph Needham and Wang Ling.

The work of social anthropologists, strongly influenced by the great pioneers Émile Durkheim, B. Malinowski and A. R. Radcliffe-Brown, reinforces the estimate already suggested of the importance of technology and techniques such as language-use for establishing societies and cultures. Various skills and devices, when they are accepted by a group, are given a place in its social structure which usually lasts unchanged for many generations. Individual members of the tribe, group or nation are born into that structure, learn its skills and use its implements, passing these to the succeeding generation before death, thus giving the group a continuing social existence distinct from the existence of the individual member. In this way the social existence of a group—its language and means of communication, customs, moral norms, legal and political institutions—are all closely related to and in many cases dependent upon its technology. For this reason, technology in a society is sometimes referred to as 'material culture' and is a vital element in man's social extension and group behaviour.

Of all the prophets of the twentieth century, one of the most provocative is the Canadian 'philosopher of technology' Marshall McLuhan. In three books—**Understanding Media**, **The Gutenburg Galaxy** and **The Medium is the Massage**—McLuhan has argued that communication technology in particular must be understood as a means of human self-extension, and as the basis for forms of society.

Man, says McLuhan, has always communicated by various

techniques; gestures, sounds, pictures or images representing things or ideas (pictograms or ideograms). Later he developed the abstract signs of written language and grammatical forms, with appropriate logical connections between the signs being accepted as conventions in a social group. He notes that the arrival of the city with its characteristic social structure and division of labour goes hand in hand with the emergence of writing. Writing and printing thus became extensions to man's basic means of communication, enabling one individual to communicate over distances and extend his influence throughout the world. At the same time the permanence of the printed word has helped to overcome the barrier of time and consolidate modern culture. Print, according to McLuhan, has made western society homogeneous, and fostered the kind of social organisations in which western man lives. It is natural for a literate man to break down processes into component parts and create jobs for each part; this is the way he has learned to think in a culture dependent upon the printed word as a medium of communication.

The life of the group, city or nation may be understood as the social extension of the self and a reflection of the media of communication most widely used. The introduction of a new medium of communication—such as television—is bound to have far-reaching effects on the way people think and organise their social lives and cooperative activities. McLuhan believes that social organisations of all kinds—political systems, civil administration, work management, educational organisations—must be understood in terms of man's social extension and means of communication. Just as various means of travel—from the horse and carriage to the intercontinental jet—are basically extensions of the legs and feet, so telescopes, microscopes and television cameras are extensions of the human eye, radio and the telephone are extensions of the human ear. Thus all communications media extend the human *sensorium*—the central nervous system—in particular ways. The form and the balance of media have a considerable influence on the way in which people think, feel, react and structure their world.

In the light of this analysis, McLuhan believes that it is not only the **content** of communications media which is important, but the **nature** of the media themselves, and the balance between them in common use. As McLuhan points out, the written word has dominated Western culture since the Renaissance, even to the point of becoming an idol in the theory of the verbal inspiration of the Bible. Likewise, reading and writing require an analytical mind and powers of sequential thought, placing these at a premium in civilised society. The use of the spoken and written word—until the arrival of television—has dominated mass communication. The danger of such concentration on the written and spoken word is that it prevents the growth of a truly general culture; it can blinker the imagination. Thus it is an inadequate extension of the sensorium, an unbalanced extension of man.

Not everyone is convinced by McLuhan's thesis. Nevertheless, he raises in a popular and provocative way the issue with which this chapter is concerned: What is technology? That question is pressed upon us with growing urgency in today's word. If it cannot be answered satisfactorily at least two major results will follow. Man will lose himself in the mere operation of his increasingly sophisticated tools, lacking any clear direction for his existence and the application of his technology. He will also be tempted to various forms of escapism and withdrawal from technical society. The tedium of a machine-operator's job, the narrow outlook of the technical specialist, the blind conformity of the organisation man points to the one fate. The signs of rejection of modern society by 'hippies', 'mystics' or 'junkies' points to the other. The slogan **Death to the Technocrats** daubed on the University of Paris in the disturbances of 1968 could indeed be the writing on the wall for our society.

7
Topics, Projects, Questions and Sources

1
Define the following terms: tool, machine, society, technique, artist, craftsman, engineer, technolgoist.

2
Debate the motion: 'In the opinion of this house, civilisation owes more to the technologist than to the philosopher'.

3
Make a series of charts to show examples of the technology of the Bronze or Iron Age, the Middle Ages, and the 1960's found in the area in which you live.

4
Prepare a ten minute talk on a technology which has been developed since 1960.

5
'The technologists of today are the magicians of tomorrow'. Discuss.

6
Prepare a group display/exhibition to indicate the relationship between art and craft or technology in one field of aesthetics.

7
What are the likely consequences of the arrival of television on ways of learning, thinking, and living together in the next decades?

8
'Technology is a fundamental expression of man's nature before God'. Discuss this statement.

9
How would you distinguish man from other animals?

Sources:

K. P. Oakley **Man the tool maker** *British Museum (H.M.S.O.)* 1952
T. I. Williams and T. K. Derry **The Shorter History of Technology** *O.U.P.* 1960
The Ages of Man (Series) *Time-Life Publishing Co.* 1967
J. D. Bernal **Science in History** (Vols 1–4) *Penguin* 1969.
Lewis Mumford **Art and Technics** *Columbia Press* 1952.
Marshall McLuhan **Understanding Media: the extension of man** *Routledge* 1964.
Sir K. Clark **Civilisation** *John Murray* 1969.
R. A. Buchanan **Technology and Social Progress** *Pergamon* 1966.
C. A. Coulson **Science, Technology and the Christian** *Epworth* 1960.
ATV Survey **Television and Religion** *Fontana* 1966.

B
Technology and Society

I
Early Tools and Societies

In 1856, the remains of a creature of similar build and bone structure to man were unearthed in the valley of the Neander, near Düsseldorf in Germany. Detailed examination of the find showed that it could be classed among the forerunners of *homo sapiens*; it became known as Neanderthal Man and was dated *c.*150,000 B.C. Subsequently, traces of the same species were discovered in Gibraltar, Belgium, France, Italy, Yugoslavia and the Crimea, and also in South Africa and Palestine. It now seems that Neanderthal man is not a direct ancestor of modern man; Java Man (*Pithecanthropus*) and Peking Man (*Sinanthropus*) are more likely to have been our immediate predecessors, although the question is by no means finally settled.

Articles found with Neanderthal Man's remains have shown that he was a tool-maker. He could make fire, and he used flint clubs and axeheads. While his teeth show that he mostly ate fruit, berries and roots, animal bones found with his remains show that he also hunted bear and wolf. He wore bone ornaments on his neck and arms. He covered his dead with ochre before binding them together for burial. In short, he lived a simple communal life with the aid of an elementary technology.

The successors of Neanderthal Man gradually extended his technology and social life. More elaborate and better made flint axes enabled him to break stronger materials. Long spears extended his reach and gave him better protection. Hunting and fishing methods were improved; ornaments such as bracelets made from bone became more plentiful; carving and building became more effective. But it was thousands of years before he organised socially to undertake the systematic manufacture of tools. Before then, early man roamed the plains in small family groups or fished from small settlements, turning his hands to every task as need arose.

His simple tools enabled early man not only to make fire, but also to devise weapons, to construct rough shelters and make simple vessels. He was able to clothe himself with animal skins. He could hunt further afield and settle in less favourable climates with these aids; he could cook food which would otherwise have been inedible: so the possibility of artificial adaptation to new environments began. At some stage in this period, more sophisticated means of communication—the origins of our language and formal written signs—enabled early man to consolidate his community life, communicate his hard-won skills and establish the beginnings of a culture.

A considerable variety of simple cultures is now known to have existed among the communities of early man in the Stone Age. Archaeological investigation in every continent has produced evidence of this wide diversity within the limits of his social life. The progression from simple to more sophisticated culture can be traced from man's tools. In the successive stages of the Paleolithic period, for example, beginning *c.* 55,000 B.C., a considerable refinement of cutting tools is observed, with increasing standardisation of form and specialisation of use. In the Early Paleolithic era in Northern Europe, knapping flints held in the hand and simple axe-heads of several shapes and sizes are found. By the late Paleolithic era, finds include axes, knives, spear-heads, bone bodkins and needles, some of which are finely decorated and all of which fall into well-defined classes of implement with easily identifiable shapes. Such progression indicates a considerable social and cultural advance which is confirmed by evidence of the use of other materials and by forms of burial and building.

As long as game was available in some abundance, hunting and fishing provided Paleolithic man and his predecessors with the basis of his diet and many essential raw materials. These were supplemented with fruit, nuts and other products of plants and trees. The evidence available suggests that during this period social life was limited to fairly small, scattered groups in which everyone participated in the constant quest for food and the necessary tasks of existence.

The cultivation of plants and the domestication of animals as sources of food and raw materials may first have taken place in North West India and the Middle East long before the fifth millenium B.C. This was a major development with far-reaching social implications. By producing his own food—whether as a herdsman or as a farmer—man was able to live in larger communities. In such expanded groups of the nomadic or agricultural type, it was no longer necessary for all the members to spend most of their time and energy getting food. A relatively plentiful supply of corn or cattle enabled time and skill to be devoted to new techniques. Weaving, rope-making, pottery and carving come from early in this stage of human development. At the same time trade became a serious proposition and settlements with permanent buildings appeared.

Food-producing man inevitably adopted a fresh attitude towards his flocks, his lands and possessions. His territory became more than a place of security where he could rear his young; it was the source of his livelihood. His cattle, land and goods were the extensions of his life and social personality. This is vividly portrayed in a number of passages in the Old Testament—particularly in the story of the wretched Achan in Joshua 7 whose family, goods and possessions were all destroyed with him as punishment for his misbehaviour. This significant change of attitude obliged food-producing man to live at a new level of social awareness, and to make basic laws for community life. The early Hebrew law codes of Exodus 20 and Deuteronomy 5 reflect this situation; they may be paralleled by the laws of earlier civilisations in the Middle East and elsewhere. In India such was the availability of food in the fifth and sixth centuries B.C. that life could be supported from fruits, nuts and vegetables supplemented by milk and other dairy products. In such a situation the teaching of Gautama Buddha that the taking of life was a sin made very good sense.

Farming requires a new set of skills, tools and techniques, and a relationship of man to the cycle of nature essentially different from that of the hunter. Agricultural life also differs markedly from nomadic life in the kind of outlook it generates as well as the skills and tools required for its successful maintenance. The mentality of the subsistence farmer cultivating his strip of land has always been different from that of the herdsman rearing sheep, goats or cattle. Here lie the roots of the age old conflict of cultures so vividly expressed in the first part of Genesis (Chapters 1–11) through the stories of the expuslion of Adam and Eve from the Garden of Eden, Cain and Abel, the Flood and Fall of Noah and the Tower of Babel. The same conflict is expressed in more popular style in the American Musical **Oklahoma** with its story of the conquest and settlement of the midwestern prairie in recent times, and its affirmation that 'the Cowman and the Farmer must be friends'.

Religion plays a central part in the culture of all known food-gathering and food-producing societies, including the life of fishing communities settled by river, lake or sea. The solidarity of the society, its relation to the world of nature and to history are invariably expressed in religious terms and celebrated in myth and ritual, prayer and cultic pattern. The village totem, shrine or church has always focused the sacred life of the settled community. Early nomadic groups carried with them cult symbols to serve the same purpose. Groups of wandering tribesmen, like the Aryans who moved into India from Iran in the second millenium B.C., tended to associate the deity with the unpredictable conditions under which they live. He was regarded as the founder of the group and its unseen protector, the Lord of its history. The religion of the farming community on the other hand was linked to the recurring cycle of nature, the round of the seasons and the rains; and to the reproductive cycles of plants, animals and men —fertilisation, birth, growth, maturity and death. Agricultural religion is closely linked to the idea of motherly sustenance, and most examples of it have a mother-goddess figure. The Earth-Mother of Northern Europe; Astarte and Isis of the Mediterranean religions; the soil-goddess of ancient China; or the earth-goddess of the Indian Khonds to whom, until very recently, a male was sacrified for her blessing on the harvest. The storm God responsible for the rains such as the Hittite Teshub or the Canaanite Baal is also a characteristic member of the family of gods, usually under the final control of El—the High God of all heaven and earth. Creation myths abound, as do seasonal festivals associated with the agricultural year. Sexual rites loom large in nature religion as members of the community link sympathetic human activity to the mysteries of nature and the myths surrounding creation and fertility. (See here **Festivals** by Gwyneth Hood, **Man and Religion Series** Part 4.)

The religion of small groups naturally expresses the needs felt by the group for protection and cohesion as well as for food and life. The High God is thought of not only as the Creator but also as the ancestral Father and as the law-giver and defender of the group. Hence in both settled and nomadic communities the earliest laws and military operations were all associated with religious rites and sanctions, and administered by leaders who were also godlings and priests.

In agricultural communities symbols and images drawn from Nature have always figured prominently, but among nomadic groups these have been less conspicuous. In the case of the nomad Israelites, such symbols were thought to be totally inadequate representations of a transcendent God. No creaturely object or idol could embody the spiritual reality of the God who acted unpredictably in history. The conflict between the cultures of the nomad Israelites and the agricultural Canaanites was never far beneath the surface of the early writings of the Old Testament. It lay behind the outbursts of the Hebrew prophets against false worship, and is reflected still today in the horror of Jew and Mohammedan at any kind of idolatory.

This tension between natural and revealed religion is a recurring element in the history of Christendom. Even the Reformation may be interpreted as a renewal of the same conflict, when faith in the God of history revived—admixed with the new individualism and nationalism of the fourteenth and fifteenth centuries in Europe—and opposed the institutionalised nature-religion under the surface of Medieval Christendom.[1] This revival of the nomadic form of faith was linked to the new wanderlust made possible by improved land and sea transport, and by new navigational devices like the compass. The conflict between a dynamic historical faith and a cyclical nature-religion is still to be detected in most societies today, and is vividly seen in the less developed nations, which are adopting the culture of industrial society. This tension of faiths certainly contributed in the past to the birth of the city.

2

The Urban Revolution

The religion of tribal communities of both nomadic and settled types provides a context for the whole of life. It offers an explanation of the universe, its origin and meaning, and of man's place and purpose within it. This is done through the mythology of the group, which is closely related to its religious ceremonies, prayers, sacrifices, dances and other customs. From early times, man's tools and technology fitted into this religious world-view, gradually influencing the mythology and religious practices through the shape they gave to work and social life, but not yet challenging man's reliance upon the unseen forces which seem to order the world and mould its destiny. Even though the bronze-using Israelites who settled in Canaan under David and Solomon had adopted the iron-age technology of the neighbouring Philistines,

[1]For a full treatment of this sociological thesis, see Peter Berger **The Social Reality of Religion** *Faber* 1969 p. 105ff.

one of the early psalmists of the time could still sing:

> *Some put their trust in chariots and some in horses,*
> *But we will remember the name of Jahweh.*
>
> **Psalm 20, 7**

In such a context it was natural for society to be regarded as divinely ordered, for kingship to be thought of as a sacred institution, and for the instruments of work—ploughs, sickles, pruning hooks—and warfare—swords, spears, chariots—to be blessed in religious ceremonies. New materials and artistic skills were devoted to the adornment of the temples, so that architecture was largely developed through religious building. In every sphere, religion dominated technology and social life.

The simple social orders of nomadic and village communities then, as now, consisted of an extended family, led by the natural father of the group, or by a man whose gifts made him the natural leader. Even in the early towns of pre-industrial Europe and in feudal society, an extension of this pattern of social organisation with similar religious forms was the norm. But at this level of social order in the period before 3000 B.C., a number of key developments occurred which stimulated the birth of the first cities.

First, new religious ideas arrived which associated the local gods with the shrines and temples in such a way that the community became willing to provide the priests or the king with the resources of food, skills and manpower which made the next stage of social organisation possible. Given a surplus of food, centrally controlled, and the necessary techniques of transport, administration, and building, it was the contribution of the new religious imagination which made the first cities.

This new religious imagination was based upon the male gods. The archaeology of the first cities of the Tigris and Euphrates valleys in Mesopotamia has shown that masculine religious symbols displaced the feminine figures which predominated in village, agricultural religion. Power thus replaced nurture as the basic notion of deity, and the noblest religious possession.

There is evidence that this change occurred when the nomadic communities from the hill country invaded the villages of the plains—in short, when the conflict of cultures modified the religious understanding of technically competent agricultural communities. It is a fact that the Mesopotamian cities of Eridu, Erech, Lagash, Kish and Ur grew up after such an invasion. The Egyptian cities, those of the Indus valley in Pakistan and the Yellow River valley in China were probably built as the city 'idea' spread to these parts. In America, the Maya, Zapotecs, Mixtecs and later the Aztecs, all built large cities when the conjunction of technical skills and religious inspiration occurred, not long after wandering cattlemen came to the farms and villages of Mexico, Yucatan and Guatemala.

Secondly, decisive factors in the growth of cities were both technical and organisational. They coincided with the discovery of metal and techniques of smelting and working it. Technically it had to be possible to farm intensively enough to produce a substantial surplus of food. This was done in the broad alluvial regions of mature river valleys with the aid of drawn ploughs and suitable irrigation systems. The early cities all developed close to such regions and they are known as hydraulic societies. There also had to be an effective social organisation, with transport and storage facilities, capable of collecting, keeping and distributing the food surplus achieved by the farmers. This organisation required full-time specialists, supported by political and military power, who could organise people to build, service and protect the city. By claiming the food-surplus for the deity, the priests of ancient Sumer in 3500 B.C. accumulated from the farmers sufficient food to support men engaged in the construction of an elaborate temple and large granaries around which the earliest city grew. In Egypt, the first cities grew up in the Nile delta before 3,000 B.C. under the Pharaohs, who claimed to represent and embody the spirit of one of the gods. From tribal beginnings, and impermanent collections of buildings, the Egyptian city was constructed around the imperial palace and the temple, with the associated grain-stores. The elaborate tombs of the Pharaohs have shown the characteristic features of early city life in the form of little wooden models: butcher's shop, barge, bakehouse and elaborately decorated tools. Unlike the Mesopotamian cities in which the population were enclosed by fortified walls, the first Egyptian cities were open settlements. Since the Pharaoh was himself a god, protection was not needed against hostile forces in the environment—and in any case the desert and the sea for the time being provided sufficient defence.

Thus religion and technology fused in the growth of early cities. New theocratic organisational forms emerged; the urban implosion began. The city became a residence of specialists, its architectural form was centred on the controlling points of its life—the religious and government buildings, and the food-stores. Systems for transport over land or water and for enforcing the political structures appeared. Wheeled carts, huge stables, ships and a permanent militia belong to this period, during which a distinct class-system emerged. The ruling élite lived centrally; around them were shops, dwellings and workrooms of the artisans—masons, potters, smiths; and on the edge of the cities lived the drivers, the carriers and the unskilled poor. A vivid picture of city life in the pre-industrial era is to be found in the story of Joseph in Egypt (Genesis 41–47) with its description of the establishment of grainstores under the Pharaohs of the Hyksos dynasties (XV to XVII).

As the late Professor V. Gordon Childe showed,[1] the 'urban revolution' was at least as important as the agricultural revolution which preceded it and the industrial revolution which followed. The city stimulated change. It became the focal point of route systems, and the place to which new ideas and inventions naturally went. Its concentration of specialists in writing, administration and crafts encouraged innovation in the realm of thought, politics and technology. Cities became the basis of national life and symbols of tradition. Some, like Jerusalem, were regarded as national sacred places—footstools of the deity. After

[1] V. G. Childe **What happened in History** *Penguin* 1967

the breakthrough from the Bronze Age to the Iron Age, new tools, weapons and means of transport made possible the extension of the idea and scope of the city further afield: nations became empires. Childe noted that urbanisation spread more rapidly in the first five centuries of the Iron Age than in all the previous fifteen centuries of the Bronze Age.

The fruits of early technological advance and of urbanisation seem to have fallen into the lap of chieftains, priests and rulers. The lot of the food-gatherers in field or pasture was not greatly advanced—at least to begin with. The life and work of the city thus offered the chance of a better material existence for the subsistence farmer in 3000 B.C., so the population of cities always tended to increase more rapidly than elsewhere.

Similar factors were at work in the great movement from the countryside to the towns in the industrial revolutions of the western world. The squalid shanty towns of southern Africa today can be paralleled by the growth of urban populations of eighteenth and nineteenth century Britain and by the grim conditions under which thousands of people lived on the outskirts of Rome, Corinth and Alexandria in the first century A.D. The difference between the present situation and that of early times lies in the size of the urban complex. Nowadays it runs into millions, whereas the early cities held only thousands of people. In the past 100 years urbanisation has proceeded at an unprecedented pace. It is expected that by the end of the century, almost the whole population of an industrialised country will live in its towns, or at least in close proximity to them.

Once the city had established itself, it contained in its own structures the seeds of further expansion. The political power at the disposal of a ruling élite presented opportunities for further extension of their rule. The cultivation of urban life generated new desires and the need for increasing the supply of materials. In particular the risk of drought, flood, pestilence or loss of food-supply through the salination of the soil, threatened famine in the city and led to strong expansionist pressures. The history of the city-states is the history of aggression and violence to achieve these ends. In this way the arrival of the city led to the idea of empire, and the rise and fall of empires is related historically to the rise and fall of major cities. This was the fate of Babylon and Susa, of Seleucia and Rameses. Athens and Constantinople (Byzantium) only avoided this pattern by attaching themselves to the succeeding empires of Rome and Islam respectively as key centres of government, culture or trade.

Empires need well-equipped armies, effective communications and good government. Those which have survived longest have been best provided with these things and with the technology and skills belonging to them. Roman roads and Roman laws survive to our own day; Roman cities lay beneath many of the medieval towns of Europe and the Mediterranean. In this great empire, the diffusion of technology, social organisation and ideas can be clearly seen.

3
The Spread of Techniques

It is impossible to be sure about the way in which different tools or technical ideas spread in the ancient world, or to what extent other civilisations developed similar skills and devices independently. There is considerable support for the diffusion theory. This explains the spread of techniques by the influence of trade, war or the settlement of conquerors from other places of origin. It was thought until recently for example, that the wheel came in this way from Mesopotamia—where it is found before 3000 B.C.—to the Mediterranean (c 2300 B.C.), and to Northern Europe (from 1400 B.C.). However, since 1950, excavations in sites in the Netherlands, Denmark, Czechoslovakia and the Caucasus have found wheeled vehicles or burials of paired oxen used for pulling carts dating from before 2000 B.C. The earliest of these are found in the Caucasus and in excavations in the plains of Mesopotamia; they are dated *c.* 3000 B.C. Professor Stuart Piggott has suggested[1] that local invention as well as diffusion may have played an important part in the widespread use of wheeled transport between 3000 and 1000 B.C.

There is plenty of evidence that people in the ancient world moved long distances by land and sea. Megalithic tombs of characteristic design are found from Britain to the Baltic and south to the Eastern Mediterranean. Obsidian—a distinctive volcanic rock—was used to make axes, knives and spearheads over large areas of Asia Minor and the Greek Islands at a considerable distance from the few sites at which it must have been quarried. The opening of the trade routes from Europe to China and India in the fourteenth century has often been presented as a singular western achievement. But for centuries before Marco Polo or Vasco da Gama the Asiatic route through Turkestan was used by Eastern traders who made their way to the Mediterranean bringing silks and spices, glass and ornaments, amber, fur and metals, and important technical ideas such as the windmill and the block press.

It was a development of a Chinese idea by Johan Gutenberg, using metal movable type instead of wooden blocks which produced the first printed books in 1454. This, following the adoption by European manufacturers of the Arabic art of paper-making, using rags, pulp and clay, began the modern history of printing, universal literacy and verbal culture. William Caxton introduced the movable type press to England in 1476 and it was immediately seized upon by religious leaders—especially the Reformers—to communicate religious ideas in homilies, tracts or translations of the Bible. The attempt by Catholic leaders to suppress the ideas of the Reformers by breaking the presses and burning the books could not succeed in halting the progress of those ideas or the new inventions which made them so widely available. Protestant Christianity undoubtedly fostered the

S. Piggott "*The Beginnings of Wheeled Transport*" in **Scientific American** July 1968 Vol. 219 No. I.

diffusion of printing technology and of literacy, and Protestant refugees like the Huguenots carried with them the skills of weaving, glass working, metal working and lacemaking.

In recent times, the efforts of Christian missionaries to communicate the Gospel to illiterate people has led to the formation of written languages and to great strides in literacy in many parts of the world. The Mission school and clinic stands alongside the Mission church, and diffusion of basic skills and basic technical ideas has proceeded hand in hand with the spread of Christian convictions and behaviour.

This process of missionary diffusion has had two major phases. The first of these occurred in the sixteenth and seventeenth centuries and was the work of the religious orders; Franciscans, Dominicans and particularly Jesuits. As a result of a visit to Japan in 1549, Francis Xavier—one of the first founder members of the Jesuit Order—concluded that it was the role of Chrstianity not to destroy indigenous cultures but to penetrate and transform them. It became the policy of the Jesuit missions to establish themselves among the upper classes and educated members of the countries to which they went in China, Japan, Korea, the Philippines and India, and to engage in service and education alongside the learned leadership of the community. In pursuance of this policy, the Jesuits passed on many western tools and techniques. A college for the liberal and administrative education of the sons of Indian chiefs was founded in Mexico in 1530; grammars were compiled of many Asian languages in the sixteenth century; clocks and maps were introduced to China in 1600, and other scientific and technical ideas were passed on in the course of study and educational work. Only a disastrous change of policy about missions to non-European cultures made by Pope Benedict XIV in 1742 prevented the continuance of this effective process of diffusion and mission.

The second main period of missionary diffusion has lasted from 1858 to the present day. 1858 marked a number of turning points: it was the year in which the British Government accepted from the East India Company the responsibility for the rule and administration of India; it marked the end of a war between China and several European nations on terms favourable to missionary activity; the Second Evangelical Revival which began earlier in the United States had just reached Great Britain and was about to stir hundreds of dedicated men and women to serve the missionary cause; David Livingstone's influential book **Missionary Travels and Researches in South Africa** had been published in 1857 stirring Christians in Britain to the challenge of the unevangelised and unexplored African continent.

From this time onwards a massive evangelistic, educational and medical effort was mounted from Europe and the United States to Africa, India and the Far East. The missionaries went out after the armies, the traders and the colonisers. Although to start with the Protestant missionaries were primarily preachers, schools were soon started in the mission compounds. From the time of Alexander Duff in Calcutta (1830–48) a college was started not only for Christian students, but also to educate all comers in the intellectual traditions of the Christian West. The Asian Christian Colleges pioneered higher education in India, South East Asia and China. Another notable Christian foundation was the American University in Beirut. Yet it is in the field of medical education and technology that the missions have contributed most to the social changes of these countries, and particularly in India. From small beginnings, two great teaching hospitals were established: at Vellore in South India, and at Ludhiana in the North. These set a pattern later to be followed by secular institutions and extended by local hospitals, clinics and dispensaries which have brought the life-saving benefits of modern medicine to millions of people. So successful has the diffusion of medical knowledge been that the present alarming growth in population of the Asian world may be largely attributed to it.

Modern missionary activity includes not only schools and hospitals, but instructional farms and colleges of technology and handicraft. This has however been only a small part of missionary endeavour until recent times, though with the help of agencies like Christian Aid it is now a major element in the Christian mission in developing countries.

While the Christian Church has contributed greatly to the diffusion of ideas and skills in these ways, and national governments have accepted its contribution gratefully for secular education, industry and welfare schemes, other religious bodies have not always welcomed these gifts from the 'wise men of the west'. These new ways challenge traditional practices and convictions, and because of this—and their association with former colonial powers—much western technology has not been adopted. The technically trained class in India is still a small minority which is widely regarded as belonging to an alien culture. Islamic leaders tend to be hostile to technological development in Egypt because of the secularism which technological advance invariably brings. In parts of Africa and India, expensive farm machinery can be seen rusting by the roadside while peasants pursue time-honoured methods of agriculture with heartbreaking inefficiency. Motive is lacking to maintain the new machines as well as the spare parts and intermediate skills necessary to do the job. With religious opposition, and the problem of finding work for vast numbers of people, old methods employing backbreaking human labour are preferred to mechanical power. The population problem, conservative religion, inadequate education and the ancient bondage to the land echoes the situation of the ages of slavery in many parts of the world.

Today the diffusion of technology takes place very rapidly through scientific and technical journals, through international trade and travel, and through the remarkable growth of secondary and higher education throughout the world. Temporary restraints on the diffusion of technical information are made for commercial reasons. Among the industrialised nations, spying for the technical secrets of rival companies is a serious problem. However lack of appropriate education is recognised to be the chief limiting factor in the assimilation of new technical ideas, but while the majority of national leaders are working to overcome this barrier, there are still those who share the attitude of mind of a pre-industrial culture. This attitude was vividly expressed by the great Indian leader Mahatma Ghandi in 1924:

> *Ideally I would rule out all machinery, even as I would reject . . . this very body, which is not helpful to salvation, and seek the absolute liberation of the soul. From that point of view, I would reject all machinery; but machines will remain because, like the body, they are inevitable.*

4
Power Before Steam

After a certain stage of technical development has been reached, the major factor in any further advance is the availability of power in the form of prime movers—basic power units. Power determines the size of object upon which a craftsman or engineer can work; it controls the dimensions of his tools and the magnitude of his machines, and consequently the output of his skill and effort. Prime movers like horses, windmills or petrol engines render energy available in concentrated form, but their characteristics of endurance, mobility and fuel supply determine the uses to which they can be put. The availability of power plays a considerable part in shaping societies—their structures, outlook and desires. The multiplicity of prime movers available to highly industrialised nations using energy from oil, gas, coal or atomic fission has made possible the structures of modern democratic society; by the variety of products of its multipowered industry, it has created the Consumer Society.

It is a remarkable feature of industrial civilisation today that there is a great similarity of gadgets, materials, techniques and systems of organisation in each developed country. Transistor radios, plastic rainwear, motor cars are much the same in London, Tokyo, Moscow, New York, Budapest, Paris and Stockholm; so is the literature, drama and television. This common culture can be attributed indirectly to the sources of power available. In particular, the development of the internal combustion engine from the early gas engines of Nicolaus Otto (1867), by Gottlieb Daimler and Karl Benz in the 1880s, has revolutionised land transport. It has made possible a completely new pattern for city structure, inland haulage and family habits.

From Neolithic times, domesticated animals like the ox have been used as sources of power. They have pulled scratch ploughs, trodden in the corn-presses, turned irrigation wheels and dragged or carried loads. The efficiency of animal power remained surprisingly low until the halter came into use. This was introduced to western Europe from the East in the late Middle Ages. Until then, animal power as a prime mover did not rate above 0.5 h.p. per unit. In any case the ox, donkey, camel or horse could not power ships or forges. For these and many other tasks, the most suitable form of power could be provided by men; moreover men were more economical to feed than oxen or horses. In mining, metal-working, building, loading and unloading and for domestic tasks, men (and women) were indispensable. To secure the power needed for the pre-industrial city, slavery was established.

The first slaves were probably prisoners of war, spared from death or mutilation to live wretched lives as hewers of wood and drawers of water. The social organisation which led to the establishment of cities, which required cheap human power, also made slaves of the members of the tribe or nation. There are concealed hints of slavery in the Israelite kingdom of David *c.* 950 B.C. The rebellion sparked off by his son Absalom was very probably fed by the popular hatred for a form of forced labour exacted by David from his subjects; it is referred to as the 'numbering' of the people (2 Samuel 24). Solomon, David's son and successor, played the part of an eastern despot. He imposed a heavy tax levy on the Israelites, and with the aid of a conscripted army used his slave labour force to build the temple at Jerusalem, his even larger palace beside it, and such important fortified towns as Megiddo and Gezer (1 Kings 9. 15–19). Solomon's story illustrates the processes by which the new nations (based on cities) organised themselves and employed human power units to achieve a new way of life. In much the same way slavery had become an essential feature of other early civilisations; in Sumer, Babylon, Egypt, the Indus Valley and later in Central America. It was also adopted extensively by the Greeks and Romans.

It is salutary to remember that the glories of Greece and Rome—architectural, intellectual, political and legal—rested on the institution of slavery. Some of the greatest achievements of the human spirit were made at the price of terrible suffering and degradation. It has been estimated that before the Peloponnesian War (431–404 B.C.) there were 125,000 slaves in Attica; of these 65,000 were in domestic employment, 50,000 were in industry and 10,000 in the mines. By the time of Cicero (106–43 B.C.) there were 200,000 slaves in Rome alone, out of a total population of one million.

Although many slaves in ancient Greece and Rome had considerable freedom and were well treated, conditions in the mines were appalling; men worked there until they died. This barbarism was justified by the needs of society and by the doctrine, in Greece, that slaves (not being citizens) did not have immortal souls. In the Roman Empire, large portions of the *ager publicus* were held by patricians as the state extended its boundaries; these lands were cultivated by slaves. Under the first Roman emperors, it was common for ordinary households to have some ten slaves apiece. As the Empire grew, the number of slaves in mining, building and workshops increased. Although some, especially in the home or school, were treated with respect, and it was possible for slaves to gain their freedom, conditions were often harsh. Rebellions of slaves occurred in 135 B.C. and in 102 B.C. in the Sicilian salt mines, and it took all the resources of Rome to crush them. Thereafter the lot of slaves was generally less easy, and the penalties for disobedience or defection were savage.

Without forced labour, Roman roads, towns and camps throughout the Empire would never have been built. The civilising benefits of the *Pax Romana* would never have penetrated to France, Spain, Germany or Britain but for the institution of human slavery and the power of human muscles taught by the Roman lash.

The religions of the Roman Empire, apart from Christianity, merely reinforced the accepted order, providing a pious escape from the boredom of life away from the battlefield.

In A.D. 313, the Emperor Constantine issued the 'Edict of Milan' following his victory outside Rome. This edict allowed all religions to be practised in the Empire. Constantine came to regard Christianity as a unifying force. He approved its teaching and played a considerable part in resolving various doctrinal crises at several Councils of Bishops. In A.D. 321, he made a further declaration establishing Sunday as the official weekly holiday in the Empire. Constantine was baptised into the Church on his death-bed in 337. Thus Christianity became the established religion of the Roman Empire. Its principles of human equality before God began to permeate society, influencing Roman Law, raising the status of women, and aiding the decline of slavery. The aggressive spirit of conquest and self-glorification was tempered, even if the Church gradually took into her structures and ceremonies some of the characteristics of the secular order. By this process of establishment, a new basis of cultural and political unity emerged on the basis of the teaching, order and discipline of the Church. But initially the great art and architecture of earlier centuries was replaced by much inferior work—not least because the wealth and manpower were not available to create it. By the fourth century, slavery had become inefficient and uneconomic and was being replaced by serfdom; the population had declined; the edges of the Empire were eroded by barbarians. The Dark Ages were beginning.

The invasions of barbarians from beyond the Alps led to the collapse first of the outposts of Empire, and then of Rome itself. When Alaric's Goths plundered the city and broke the political structure of the ancient world, Augustine, Bishop of Hippo, was moved to write **The City of God**, a view of human history and society in the light of God's eternal kingdom, which was to prove of enormous influence in the later Middle Ages. Providentially, the Christian Church preserved the law and government, rescued much of the ancient intellectual tradition in her monastic schools and libraries, and did much to preserve order and communication even though the substance of the Empire had been badly shaken. Such was the strength of the Church that in the course of only four centuries, her teaching and practices were to transform the pagan cultures of the whole of Northern Europe and Russia.

The power unit which was to replace the lost energy of slaves was developed only slowly. It was the water-mill. A Norse mill is known to have existed in the first century B.C. Simple horizontal mills like this appeared in several parts of Europe, in the Mediterranean and in China, up to the fourth century A.D. They were used for grinding corn. Roman inflexibility found little else for mill-power, and it was not until the tenth century that water mills were used to any great extent for other purposes.

The vertical mill, invented by the Roman engineer Vitruvius as far back as the later years of the first century B.C., gradually replaced the horizontal mill in the tenth and eleventh centuries. This mill employed gears of the kind suggested by the Greeks. By the tenth century iron-casting was sufficiently advanced for reliable gears to be manufactured. Whereas the horizontal (Norse) mill has a power output of approximately 0.5 h.p., the vertical (Vitruvian) mill rated at approximately 3.0 h.p., and was thus a real advance on animal power units like the ox or man himself. The efficiency of the vertical mill was steadily improved during the Middle Ages to give a maximum output of 60 h.p.—a good example of necessity being the mother of invention. Since these mills required fast-flowing streams of water, they had to be sited in hilly country with a fairly high rainfall. The transfer of the English textile industry from the south of the country to

Yorkshire and Lancashire in late Medieval times was a direct result of the development of the water-mill and the availability of fast streams flowing off the Pennines.

In the eleventh century, the windmill—originally a Chinese invention imported into the West—was widely adopted as an alternative source of power. It produced a similar output to that of the watermill and could be used in open country exposed to the prevailing winds, in Anglesey, West Wales, East Anglia and the Low Countries. Indeed, so many windmills were built throughout northern and western Europe that the Pope introduced a windmill tax! For centuries thereafter—until the arrival of the steam engine—these were the major prime movers.

The siting of static power units like the watermill determined the growing points of industrial communities in the Middle Ages. The **Domesday Book** (1068) lists over 5,600 watermills for about 3,000 communities in England; one for every 400 people. The conjunction of such power units with the availability of metallic ores, trade routes and skilled craftsmen determined the location of the Medieval towns and shaped the economic map of the Middle Ages. In these towns ingenious men, spurred on by ambitious merchants, enlarged and improved their machines, finding more and more tasks for them to perform; fulling broadcloth, grinding metallic ore, and hammering iron, as well as grinding corn. Although the preoccupations of the Church and the trend of scholarship lay elsewhere, it was in these towns that the capitalist system and modern banking developed, making possible the competitive enterprise which was to give Europe mastery of the world through her technical ability, military power and the aggressive search for wealth In spite of terrible setbacks like the bubonic and pneumonic plagues and the Hundred Years War, medieval merchants, craftsmen and engineers laid the foundations of the western world as we know it. Just as surely the artists, carvers and architects of those days shaped the significant buildings and standards of culture for centuries to come. All this took place long before the modern scientific movement began in the seventeenth century, and long before the authority of Aristotle or Ptolemy or the Pope was attacked directly.

Alongside static units of power, the Middle Ages harnessed new animal energy from the horse. The invention of the stirrup had made this creature a major factor in warfare, and firmly established a socio-economic system designed to maintain mounted knights and their supporting cavalry, which we know as the Feudal system. More important in the long term, however, was the use of the horse-collar and the horse-shoe. The former may well have been an idea imported from the East where it is known to have been in use some centuries earlier. As a power unit, the harnessed horse was mobile; its use dramatically improved both the methods and yields of farming and means of transport. More than any other factor the harnessing of the horse made possible the establishment of nation states in Europe, providing the economic resources, the means of political control through a mobile military force and the communications which national unity required. At the same time, the development of the sailing-ship and new navigational aids made possible the growth of the mercantile empires and brought home to Europe the wealth and resources of the rest of the world for further industrial advance.

One of the major contributions of Christianity to the rise of industrial Europe was its emphasis upon work. Many classical thinkers such as Plato, Aristotle and Cicero stressed the importance of leisure and intellectual activity over against physical labour and craftsmanship. The Christian tradition on the other hand, taught that work was good, necessary and godly. Man was a fellow-worker with God. The Holy Spirit was given to craftsmen as well as intellectuals. Jesus was the son of a carpenter. Like the Jews, Christians denounced idleness and praised the dignity of work, stressing its necessity for individual and social health.

This theme was taken up by the religious orders based upon the **Rule of St Benedict of Nursia** (*c.* 540), the founder of western monasticism. The Rule insisted that monks should spend their time on manual labour when they were not engaged in worship, divine reading or meditation. Converts to Christianity had to assimilate the Christian view of work, and by the Middle Ages, the gospel of work was deeply engrained in the minds of the common people. This emphasis was to persist and grow through the period of the Reformation, notwithstanding the troubadour tradition—expressing a romantic view of life in song and poetry—and the rise of intellectualism in the wake of the new learning. It has come down to us today in a secularised form, making the use of leisure a difficult task for older people.

The Church's teaching about the dignity of work was linked to its doctrines of social order, and the rights and privileges of the individual according to his place in it:

God bless the squire and his relations,
And keep us in our proper stations.

The Church also taught the notion of a 'just price'—that goods should be priced not simply on the basis of supply and demand, but in relation to the needs of society. The idea was based on the maxim 'Do unto others as you would they should do unto you'. Price-fixing for basic commodities and manufactured goods was enforced by the moral authority—and occasionally the political pressure—of the Church, together with the practical authorities of cities and the guilds. In a curious way the same principle returned in the United Kingdom with the idea of the Prices and Incomes Board!

The Church also taught that usury was sinful; that is to say that it was considered wrong to take interest on capital loans. The Scriptural prohibition was reinforced by the practical desire to prevent men from living without working. Soon such a strict rule had to be bent to allow for the growth of industry. Large-scale trade and manufacture could not proceed without capital sums of money, and capital was not to be had without the payment of interest. The Church's prohibition helped to delay the growth of industry, and for long put the Jew in the unique role of the moneylender in the Middle Ages.

With the dissolution of the Medieval world and the upheavals of the Reformation and European nationalism, the restraints upon capitalism were removed and commercial expansion began in earnest. The Protestant ethic also stressed the doctrine of work, and fostered the notion of individual enterprise and national achievement. Merchants and craftsmen in the growing towns became the mainstay of the new nations. They were the educated, loyal, bourgeois middle class, nurtured on thrift, goaded by the prospect of wealth and inspired by the gospel of work. The paintings of Rembrandt display and offer a telling comment on the bourgeoisie of this period.

The human power unit was never outmoded in this long development, and is certainly not redundant even today. The advent of other forms of prime mover made possible new processes and methods which could not otherwise have arisen; but in many areas, particularly mining and farming, human labour was as essential as it had always been. In these two activities in particular, the rise of the capitalist system and the accompanying economic doctrine of mercantilism led to new forms of slavery, Christianity notwithstanding. Conditions in mines had always been poor, even for independent miners like those of the Forest of Dean who worked for themselves. The coming of capitalism to areas like Tyneside in the seventeenth century meant that deeper pits could be made and more labourers employed. The wealthy landowners of the district provided the capital, and bought the labour for a meagre sum. By 1700 large numbers of men, women and children were working full-time in the pits to provide both for the expanding markets for coal, and the profits of the owners.

However, the most appalling story of western exploitation of human labour is not to be read in the history of mining, where at least the workers had a wage and a modicum of freedom. It was in the slave-trade which supplied labour for the plantations of the New World.

Slavery on a scale unheard of even in the ancient world began in the new era of merchant venturing, as a direct result of the growth of capitalism in Europe. Slaves first went to the New

World soon after the first colonists. There were Negroes on Haiti as early as 1501. The main cargoes of slaves started to arrive in the West Indies from West Africa in 1518, and the traffic continued until the 1870's, although the trade was largely suppressed after 1865. Nobody knows how many slaves were taken across the Atlantic. Conservative estimates suggest that the number exceeded fifteen millions, and upwards of nine million more died on the way. It is true that some of the white colonists had also been forcibly transported to the New World, but they were freed after a few years. Negroes imported after the seventeenth century were chattels for life, condemned with their descendants to perpetual servitude. It became cheaper for West Indian planters to buy them than to breed from them.

The story of the slave trade implicates nearly every European power except the Italians, the Austrians, the Poles and the Russians, who had no colonies. It also implicates both the Arab traders and the coastal tribes of East and West Africa, who caught and sold Negroes from inland tribes to the slave ship captains. It implicates the reckless seamen who made their fortunes from the business; men like Sir John Hawkins, Captain Billy Boates, Francisco Feliz da Sousa, the prince of slavers, and Captain (later the Reverend) John Newton. Newton later wrote of his voyages on the slave business that 'it renders most of those who are engaged in it too indifferent to the sufferings of their fellow creatures', and treating Negroes like cattle 'gradually brings a numbness upon the heart'. Most of those concerned were physically ruined or personally degraded beyond redemption; it was the businessmen of ports like Liverpool, Bristol, Amsterdam and Lisbon who made the profit.

Nevertheless it was the English who led the anti-slavery movement and who did most to abolish the trade. Private and public doubts were first expressed in the American colonies of Massachusetts in 1641 among lawyers and Quakers. Throughout the eighteenth century, measures were passed in the Northern Colonies to restrict the trade, and protests were made to the British Government. John Woolman the Quaker, John Adams,

Granville Sharp as well as Thomas Clarkson and William Wilberforce are to be remembered among those who strove to abolish the iniquities of slavery in the face of violence, public hostility and parliamentary inertia.

In the event, President Jefferson signed a bill to abolish the trade on 2nd March, 1807, just before the abolition law was enacted in Britain, but it took many years and many lives to stop the smuggling of slaves to Georgia and Louisiana. The American Civil War had to be fought to gain some release for slaves in the Southern states. Only the acceptance by the western nations of the principles of liberty and justice, supported by a new Christian humanism, brought this horrid business to an end. Paradoxically, those who still deny the black man his full dignity in South Africa, or in the United States or Great Britain usually do so on the basis of a biblical justification for the social conditions that existed in 800 B.C. In this way white people gloss over a deep-rooted fear and show themselves reluctant to change a socio-economic system founded upon the power available from black muscles. Until new technical devices arrive to do black men's work, or a reformation of religious ideas occurs, it is doubtful if that attitude will change.

5
Steam, Steel and Slums

Two of the most inventive minds in recent European history were those of Leonardo da Vinci (1452–1591) and Isambard Kingdom Brunel (1806–1859); they lived at the beginning and the end respectively of the first major phase of invention and industrialisation in modern Europe. We are now living in the middle of the second phase. Leonardo's work points to the great needs of inventors at the start of the period: strong, workable materials for the construction of new machines, and adequate sources of power to work them. The majority of his engines remained sketches on the drawing board because these needs had not been met in his time. Brunel's achievements—his bridges and tunnels and ships (like the **Great Britain** steamer)—employed to the limit the iron and steel available in mid-Victorian England, and the steam power from coal harnessed by the engines of Watt and Stephenson. Like Leonardo, Brunel was an artist, and monuments to his design like the Clifton Suspension Bridge show how man-made constructions could achieve beauty and dignity in the midst of much that was either badly designed or badly built, for example, the ill-fated Tay Bridge.

The recent history of Europe can be written in terms of the power and materials revolution. These developments were fundamental to the British industrial revolution, and in turn led to the expansion of the towns and cities in the nineteenth and twentieth centuries.

After the middle of the sixteenth century, the demand for power outran the capacity of watermills and windmills. From 1550 onwards, serious investigation began into the properties of steam as a source of power, though satisfactory engines did not appear until much later. At the same time, the smelting of iron ore increased to provide cast iron for more and larger machines. Mining and foundry-working became the first industries to adopt the capitalist system in Elizabethan times, money being provided by one man or by a group of owners, while work was carried out by employees for wages.

In order to meet this demand for fuel and metal, mines were dug deeper. Coal mines reached a depth of 400 feet in 1700, 600 feet in 1750; and the problem of pumping out water became acute. The steam engines of Thomas Savery (**The Miner's Friend**, 1698) and of the Marquess of Worcester (1663) were attempts to meet this problem, but they were highly inefficient. Large numbers of horses continued to be used until Thomas Newcomen combined a piston and cylinder with valves and a steam boiler successfully in 1712. Although it used large quantities of fuel, Newcomen's engine really worked and was adopted for use in Cornish tin mines and in the coalmines of Northern

England. Engines of this type using a piston in a cylinder to operate a beam-acting pump were still in use as late as 1830.

At this time, serious attempts were being made to improve the efficiency of watermills and windmills. John Smeaton (1724–92) the famous engineer, engaged in intensive laboratory work. He introduced the breast water wheel, and made important improvements to windmills. He then turned his attention to discover the most efficient form of Newcomen steam engine. Although Newcomen engines could not drive rotary machinery, they could pump water. The firm of Abraham Darby, ironfounders at Coalbrookdale in Shropshire, used such an engine to pump water to drive ten water wheels in 1732.

It was James Watt who took the major step in improving Newcomen's engine, ensuring that the steam did not condense in the cylinder, thus dramatically increasing its efficiency. Then, in partnership with Matthew Boulton of Birmingham, he had cylinders carefully bored and other parts accurately made in Boulton's workshops in Soho. The result was a vastly improved pumping engine, shortly to be followed in 1781 by rotary action and in 1782 by the double acting piston engine. These low pressure engines were quickly adopted by British industry. By 1800 there were eighty-four Watt engines in cotton mills, thirty in collieries, twenty-two in copper mines, nine in woollen mills and others in rolling mills, forges and foundries. These engines met the demand for power, and immediately gave British industry a lead over its competitors in the markets of the world. Another Englishman, Richard Trevithick, made the next development of high-pressure, non-condensing engines early in the nineteenth century, and these eventually replaced the Watt low-pressure engines in industrial use.

The application of steam engines to locomotion on land and sea is a familiar story. Robert Fulton—the much travelled American engineer—built the **Clermont** in 1807 and temporarily the U.S.A. held the lead in steam navigation, but when the paddle steamer **Great Western** crossed the Atlantic in 15 days in 1838, inaugurating the first regular transatlantic service, Brunel brought the lead firmly back to Great Britain. Brunel went on to build screw-driven ships: the **Great Britain** which crossed the Atlantic in 1843, and the huge **Great Eastern**.

James Watt considered the idea of using a moving steam engine on rails impractical, on the grounds that the wheels of the engine would not grip the track sufficiently to enable it to haul worthwhile loads. With the work of Richard Trevithick on high-pressure steam engines this problem was overcome, and several demonstration models were made after 1804. William Hedley of Newcastle incorporated the idea of a 'steam blast' into his locomotive in 1813 (**The Puffing Billy**) using steam from the cylinders to improve the blast for the fire. The locomotive was used at the Wylam Colliery near Newcastle as a short-haul engine between pithead and canal. George Stephenson was born at Wylam in 1781. He became surveyor and engineer for the Stockton and Darlington Railway connecting Witton Park Colliery and the sea, with a branch line to Darlington. His first locomotive, the **Locomotion**, was used on the line when it opened in 1825. Stephenson then went to the Liverpool and Manchester Railway for which he and his son Robert designed the **Rocket**. The railway opened in 1830 with 35 miles of track and was used for passengers and freight. The Railway Age had begun. By 1849 there were 5,000 miles of track in Britain. By 1850 all the main towns were 'on the railway' and a new, rapid, transport system had united the country—or at least those areas of the nation which shared the same gauge of track! The effect of the railways in uniting the nation, spreading people and ideas as well as goods and raw materials, was immediate. The steam train caught the Englishman's imagination and penetrated his art in pictures like J. M. W. Turner's **Rain, Steam and Speed** (1884), though the effect of railway dirt on living conditions was more accurately depicted in the drawings of the French artist Gustave Doré in 1871 and by Charles Dickens in **Dombey and Son**. The limiting technical factor determining the size of an effective social or industrial unit had overnight been altered by the coming of the railways with their snorting monsters of steel.

To meet the demand for iron and steel, the scale of mining and traditional forms of smelting with charcoal had to expand rapidly, to the further devastation of the English forests. Abraham Darby successfully smelted iron with coke made from anthracite coal in 1709, but he needed an adequate blast to do this. Until the Boulton and Watt engine was available at the end of the eighteenth century, the majority of furnaces continued to use charcoal, and the industry was located near woodlands. Moreover, the cost of British iron was higher than better-quality Swedish iron. John Wilkinson, the ironmaster of Bradley, installed the first Watt engine for a coke-fired blast-furnace in 1760. Thereafter the iron industry was linked to the coalfields.

To start with, up to 10 tons of coal were needed to produce one ton of pig-iron. In 1843, Henry Cort patented a puddling and rolling process which allowed the large-scale production of cheap wrought-iron from the impure British pig-iron produced with coke. The steel industry of South Wales then began, using home-produced wrought-iron rather than the imported Swedish product. The subsequent introduction of the steam hammer and the steam-driven rolling mill completed a series of major innovations which strengthened the British industry and united it around the coalfields. The scale of production of pig iron was increased by two new processes: James Nielson introduced the hot-air blast to cut coke consumption and increase the size of the furnaces in 1829, and Henry Bessemer patented his new steel-making process in 1856. The vast operation of modern steel-making thus began, requiring a huge capital outlay, and the revolution in the industry and its structure was complete until the recent innovations of automated, oxygen-blast, open-hearth furnaces arrived. It is the considered judgement of many economic historians that it was the iron and steel industry, even more than the cotton industry, which played the most significant role in the process of British industrialisation.

If iron and steel making were the great capitalist industries, it was nevertheless cotton which effectively led the factory system in Britain. The manufacture of textiles had always been a major part of British economy, long before the industrial revolution. Woollen cloth produced from the fleeces of sheep on the English pastures and woven in the cottages of Cotswold, Yorkshire or East Anglian villages was renowned in cool latitudes. Children cleaned and carded the wool; women span the yarn; men wove the broadcloth. The pre-industrial cotton industry was likewise a household affair in the north west. For many homes it was subsidiary to farming. But cotton was a new industry in the eighteenth century, and therefore more ready to adopt new machines. It faced production problems, since three or four spinners were needed to supply one weaver. Further, the demand for cotton was strong at home and overseas. The domestic population was increasing and incomes were rising. In this way the scene was set for innovation.

A series of inventions transformed the cotton industry inside fifty years into the first fully mechanised production system in the world. Kay's flying shuttle (1730) was widely adopted by the weavers after 1750; Paul's carding machine (1748) was in general use by 1760. Hargreave's spinning jenny—patented in 1770—found its way into many homes as well as the new factories, thus relieving the bottleneck in the spinning process. Arkwright's water-frame (1769) was a factory machine which produced yarn strong enough to be used as warp as well as weft; this led to the marketing of a new cotton cloth without a linen warp. Crompton's mule (1779) combined the principles of the spinning jenny and the water-frame and produced a finer yarn. Then, with the appearance of the Boulton and Watt steam engine in 1785—the same year that Arkwright's protective patent ran out—the industry was finally transformed into a large, factory-based, mechanised system. Between 1780 and 1800, the import of raw cotton from America increased eightfold. By 1815, exports of cotton textiles totalled 40% of the value of domestic exports, compared with 18% for woollen goods. Prices fell, quality rose, exports doubled and redoubled to markets all over the world. The factory system, with extensive mechanisation proved its worth to a society geared to the acquisition of wealth.

Many factors aided the industrial revolution in Britain. A number of recent studies have attempted to isolate the most significant of these in order to understand better the processes of economic growth and industrial 'take-off'. Such knowledge could be of value to developing countries today in planning for a similar industrial revolution. Professor W. W. Rostow has argued that the development of the cotton industry was the decisive step which paved the way for the factory system and mechanisation. Professor J. U. Nef, on the other hand, held that the key to the British industrial revolution lay in the first capitalist undertakings of the sixteenth century, when large scale coal-mining, iron-making and woolweaving industry began. But many other factors have to be taken into account as well; the improvements in agriculture of the eighteenth century and a remarkable series of good harvests between 1740 and 1760; the steady rise in population after 1740, aided by years of internal peace and the gradual disappearance of the black rat whose fleas brought the deadly bubonic plague; the availability of raw materials such as wood, coal, iron-ore at home and raw cotton from the colonies.

To these material causes must be added a desire on the part of people in Britain to gain wealth and to enjoy life at a new level, unhampered by many ecclesiastical warnings about the perils of riches and covetousness. This new attitude also significantly encouraged innovation and the rise of modern science (see **Religion and the Sciences** pp 21–27). The break-up of the medieval World-view projected by the Church, and the disenchantment of the world from mystery, miracle and magic, led to a transformation of attitudes and a new stress on practical, material preoccupations. The prohibition on usury was lifted, and spiritual concerns were separated from much in ordinary life except honesty, hard work, thrift and individual salvation. Behind the social, geographical and technical factors, this underlying shift of convictions must be reckoned as a decisive factor in the British industrial revolution.

The close association of the rise of Protestantism—especially in its Calvinistic form—with the growth of private capital led Karl

Marx (1818–83) to the conclusion that the religious ideas of the Reformation were no more than the spiritual rationalisation of the facts of life in a capitalist economy. The separation between owner and worker, the class system, and the exploitation of poor by the rich, were apparently justified by this form of religion. If, as Calvin had stressed, man was a sinful, undeserving wretch, utterly dependent on 'grace', then the Christian system was simply a spiritualised version of the workers situation *vis-à-vis* the capitalist. Borrowing a phrase from Charles Kingsley, Marx proclaimed that 'religion is an opiate of the people'. He insisted that the 'criticism of religion is the beginning of all criticism'.

A different view was taken by the sociologist Max Weber (1864–1920). He argued that capitalism certainly emerged under the influence of Calvinism, but that the religious ideas of the Reformation were not necessarily a reflection of the facts of economic life. Weber considered that religious ideas could be regarded as real factors in the growth of industrial society. His thesis was popularised by later writers such as R. H. Tawney in his book **Religion and the Rise of Capitalism**. The representative figure of religious influence on social and technical change in industrial England is Joseph Priestley (1733–1804). Priestley was the son of a Yorkshire cloth-dresser who belonged to the Independents. He was educated at one of the famous Dissenting Academies in the Calvinistic faith and a wide-ranging, open-minded intellectual tradition. In addition to his well-known researches in chemistry and physics, for which he was made a Fellow of the Royal Society, Priestley taught history, theology, philosophy and politics, first at the Warrington Academy, and then in Leeds and Birmingham. As a Socinian—a follower of the rationalist sect of Socinus, which gave rise to modern Unitarianism—Priestley believed enthusiastically in Necessity, Progress and Perfectability. In his **Essay on Government** (1768) he wrote:

> *The great instrument in the hand of divine providence of this progress of society towards improvement* **is** *society and consequently government. All knowledge will be subdivided and extended; and knowledge, as Lord Bacon observes, being power, the human powers will in fact be increased; nature, including both its materials and its laws, will be more at our command; men will make their situation in the world abundantly more easy and comfortable; they will prolong their existence in it, and will grow daily more happy, each in himself, and more able to communicate happiness to others. Thus whatever was the beginning of this world, the end will be glorious and paradisaical, beyond what our imagination can now conceive.*

Alongside the religious shift of emphasis, and mixed with it, another major influence was the growth of individualism, usually tinged with moralistic notions of the virtue of work and wealth, and the vices of laziness and poverty. Man now stood on his own feet before God, his fellows, and the world of nature. In Britain this attitude was given political expansion by the Whig Party.

It is sometimes suggested that conditions of work in mines, factories, and the industrial towns which mushroomed almost overnight about them, were suddenly and unbelievably appalling. It is usually forgotten that the lot of the cottager at his spinning and weaving, and of the poor who worked on the land, were generally rather worse. Children worked in the field and in the cottage at least as long hours as those in pit or factory. Women's place in pre-industrial England has little to commend it to emancipated modern thought. Life for all the family was nasty, brutish and short. The infant mortality rate sometimes reached 25%, and life expectancy was no better than in undeveloped countries today—that is, around thirty-five to forty years. It is small wonder that the opportunity of earning more money tempted thousands to the mines, foundries and factories, even though they lost some of their independence and cast their luck on an untried commercial system. No doubt the rewards were great enough to justify the different squalor of the town and the new risks of work in pit and factory out of the weather.

In the large-scale factory system pioneered by 'King Cotton', one decisive change was the new relationship between worker and owner. The direct, personal link was replaced by the rule of the overseer or foreman. The worker became a cog in the producing machine, his—or her—life was geared to the demand of the system. The easier rhythm of cottage or farm work was replaced by the relentless pace of the factory. It is small wonder that human beings reacted to the new strains imposed by these conditions by turning to drink, debauchery or religion as a way of escape. Peter Gaskell, the Cheshire surgeon, wrote in 1833 in his book **The Manufacturing Population of England** of the new master cotton-spinners and weavers of Lancashire:

> *These were men, who had sprung from the ranks of the labourers, or from a grade just removed above these—uneducated—of coarse habits—sensual in their enjoyments—partaking of the rude revelry of their dependants—overwhelmed by success—but yet, paradoxically as it may sound, industrious men, and active and far-sighted tradesmen.*

And of the houses of the labourers themselves Gaskell had this to say:

> *The houses of great numbers of the labouring community in the manufacturing districts present many of the traces of savage life. Filthy, unfurnished, deprived of all the accessories to decency and comfort. What little furniture is found in them is of the rudest and most common sort, and very often in fragments—one or two rush-bottomed chairs, a deal table, a few stools, broken earthenware, one or more kettles and cans . . . a bedstead or not as the case may happen to be, blankets and sheets in the strict meaning of the words unknown—their place often being made up of sacking, a heap of flocks or a bundle of straw . . . and all in a single room, which serves a place for domestic and household occupations.*

> *Many of the houses are totally undrained or only partially. The whole of the washings and filth from these consequently are thrown into the front or back street, which being often unpaved and cut up into deep ruts allows them to collect into stinking and stagnant pools; while fifty, or more even than that number, having only a single convenience common to them all, it is in a very short time completely choked up with excrementitious matter. The demoralizing effects of the utter absence of social and domestic privacy must be seen before they can be thoroughly understood, or their extent appreciated. By laying bare all the wants and actions of the sexes, it strips them of outward regard for decency . . . they do not scruple to commit acts in the presence of each other, which even the savage hides from the eyes of his fellows. . . .*

The impression is confirmed by official reports, as well as by the writings of Charles Dickens and Frederick Engels.

The indifference of employers to the lot of their workers was a longstanding feature of social life rather than a new phenomenon in the factories. What the new processes did introduce was the impersonal intermediate relation between worker and overseer in place of the direct relation between the labourer and his master: the content of the relationship was little altered.

The conditions of the working class were slowly changed by a combination of political agitation, the stirring of conscience among the middle and upper classes which resulted in more human laws, and the introduction of new and more efficient machines and methods. But at the same time the processes of industrialisation brought new risks to health: silicosis in the mines, typhus (factory fever), grinders' asthma, phosphorus necrosis of the jaw, tuberculosis (consumption), as well as death or injury from blasting, pitfalls, unprotected machinery, fire or burns. It is worth noting that compensation for industrial injuries was only introduced in Britain in 1880.

The beginnings of the industrial revolution in Britain do not generally present a pleasant picture. Noteworthy efforts to provide better conditions in homes and factories were made by the pioneer socialist Robert Owen (1771–1858) of New Lanark, and by other employers like Josiah Wedgwood, Samuel Oldknow of Mellor and the Peel family, but these must be counted as exceptions to the general rule. The legacy of tenement buildings, unplanned towns, lack of space and privacy for workers, and unnatural rhythms of work in industry are still with us.

On the other hand, personal living standards have risen progressively as a result of industrialisation, in spite of setbacks like world wars, slumps, and quickening commercial competition from abroad. Standards of diet, hygiene and education have been raised dramatically in the last century and a half. Public administration and social services of an entirely new level have come into existence. Political emancipation of workers and women has followed the adoption of the new social system required by factory and commercial work. In spite of the heavy cost in terms of human life and a ruined environment, a new humanism has emerged in this period which is reflected in the present provisions for social welfare, the administration of the law, and personal concern for those in need. The clock will never be turned back in industrialised countries, and other nations now consider it their equal right to pay the price and share the benefits of industrialisation. Man's ambiguous relation to history—being made by it as well as making it himself—is once more exposed in the story of every industrial revolution.

6

The New Society

It has been suggested above that the first stage of industrialisation is still being worked out. Mechanisation is still being introduced in certain industries such as farming and market gardening, road, rail and ship haulage, and in the home. But already a second phase has begun, bringing automatic processes of mass-production and sophisticated techniques of organisation and control, and using a wide range of new materials from plastic fabrics to carbon fibres. These developments are leading to a new society, different in many ways from the class-structured social order of Victorian times.

One major element in the rapid rise of productivity, and hence of standards of living and relative wealth of industrial states, has been the introduction of mass-production methods. As with many modern techniques, the basic principles have been known and used for many years. The mass production of crude articles such as headed pins began in Nüremburg in 1680. At a factory owned by Christopher Polhem in Sweden, 100 men were employed on machines driven by water power for the manufacture of tin-plates, nails, hammer-heads and metal pots. This was in 1700. In England, Marc Brunel and Samuel Bentham set up a factory in 1808 to manufacture pulley blocks for the Admiralty, using machinery designed and made by the tool-maker Henry Maudslay. Ten unskilled men operated forty-four machines, replacing 110 men using older methods, and achieved a production six times that of any other factory. Early mass-production gave a roughly similar article. The products of Brunel and Bentham were standardised so that the parts could fit together reasonably well. The improvement of standards of precision to make possible the interchange of mass-produced parts of an article was the crucial next step. It occurred in the gunmaking industry, first—unsuccessfully—in France, and then in Eli Whitney's musket factory in the United States in 1880. With this new system of production came new precision tools: the Whitney milling machine in 1818; the Blanchard lathe in 1819; the Root drop-hammer in 1855; and the automatic lathe in 1862.

By 1860, mass-production techniques were well established, and were employed in the manufacture of the Singer sewing-machine, the mechanical plough and later, the motor car. R. E. Olds introduced a small mass-produced car to the market in 1902 and in 1905 turned out 5,000 of them. Henry Ford marketed 10,000 cars in 1909 and introduced the conveyor-belt system of assembly in his factories. Today, automatic computer-controlled systems are being installed in the motor industry and many other manufacturing plants. Fewer operatives are required to achieve a much increased production, although highly qualified maintenance engineers are needed to service the machines and controlling devices. Given the skilled manpower for these tasks, all that is required for a huge leap in productivity is the necessary machinery and controls, specially designed for each operation; the large capital sums to purchase such plant; and of course the consumer outlets for the products at home or anywhere else in the world. Industries in which these new machines are being used include printing, the chemical and oil industries, food-preserving and packaging and machine-tool manufacture. This new step forward is certain to accelerate rapidly in other sectors of industry as smaller, cheaper automatic controlling devices become available. The implications of this development for industrial training, the availability of new materials and gadgets, and for social life in general are enormous.

In administrative processes as in manufacturing techniques, there are signs of revolutionary changes. In place of the vast quantities of paper used for records in commerce and government transactions, simple rationalised dockets and electronic information storage and retrieval systems are being introduced. Careful analysis of organisation and business methods has led enterprising firms such as Marks and Spencer to considerable economies in costs for administrative materials and manpower: a trend which contradicts Parkinson's Law. Government departments and large commercial concerns are also pioneering new electronic systems. At the same time, the use of the telephone is steadily increasing, and videophone links are likely to be generally available in a short time. The extensive use of teleprinters is another application of modern technology to business communications. These changes mark a considerable departure from the rows of clerks entering information in copper-plate script on massive ledgers, so often found in business houses of the nineteenth century.

Most of the office-staff in that period were men. Nowadays a majority of secretarial employees are women, and women are found in large numbers in new careers like that of computer programming, market research, and personnel management. The invention of the sewing machine by Singer in 1850, the telephone by Alexander Graham Bell in 1876, and the typewriter by C. L. Scholes, an American printer, and produced commercially by the Remington Company in 1873, opened the way to new careers for women in commerce and government, just as Florence Nightingale and the worthy Miss Beale and Miss Buss did in the professions of nursing and teaching respectively. There were 7,000 women in office employment in Britain in 1881, 146,000 in 1911 and over 500,000 in 1951. The status of women in society has been considerably altered by these opportunities, and also by the vital role they played in industry, agriculture and the social services in the 1914–18 and 1939–45 wars. Middle-class inhibitions about women at work—inherited with some justification from the previous century, when the lot of women workers in the factory was to be exploited for cheap labour and male licence—have been steadily overcome. Today, a substantial part of the commercial, industrial and governmental labour-force consists of women in full- or part-time employment who now enjoy the prospect of equal pay for equal work with men. Thus the opportunities afforded by new inventions, together with rapidly changing social attitudes and educational facilities, have all combined to restructure this aspect of the social scene in a remarkably short time.

The inevitable difficulties facing women who attempt the double roles of home-maker and worker have been eased by the availability of new machines for the home like the washing machine, the vacuum cleaner and the refrigerator, and by the increasing domestication of the menfolk of the household. This has led some observers to suggest that the new society is becoming excessively home-centred rather than community based, and increasingly dominated by women and their needs. The demand for nursery schools, welfare services for families and old people, and for the second motor car point in this direction. But few men nowadays wish their wives to be tied to the kitchen-sink—if only for the improved standard of living which results from their going to work.

The first phase of the industrial revolution, from 1740 onwards, was marked by a sharp increase in the growth of the population. At the same time, the population of the towns began to rise rapidly. Between 1851 and 1901 the major industrial cities in England doubled their population: Birmingham grew from 233,000 inhabitants to 523,000; Sheffield from 135,000 to 407,000; Manchester from 303,000 to 645,000; and London from 2,500,000 to 4,500,000. By the turn of the century, Britain had become the first urbanised nation. Today, all industrial nations are highly urbanised, and the process of expanding towns is accelerating rapidly throughout the world. The rate of growth of cities over 100,000 inhabitants in the period between 1950 and 1960 was twice that of the preceding 50 years. By 1960, the proportion of the world's population living in cities had reached 33%. It is predicted that by the year A.D. 2000 some 90% of the world's population will live in cities, while the total population of the planet will have reached 7 billion persons—although such predictions are notoriously unreliable and generally far too low.

The latest phase of technological development has had other important effects. For example, it has made possible planned towns and rapid rebuilding. The planning movement is centuries old, going back to Mohenjo-davo in the Indus valley in 3000 B.C. It was revived after the 1914–18 war through the Garden City movement, demonstrated at Letchworth, Port Sunlight and Bournville. After 1945, town planning became a serious responsibility of central and local government. The New Towns Act (1946) and the Town and Country Planning Act (1947) made a start with the urgent problem of rehousing and land use in Britain. Much remains to be done, but with more study, practical experience and public awareness of the problems, more and better planning is creating attractive new towns and reshaping older cities with increasing rapidity. The availability of new materials, techniques and tools for excavation, building and prefabrication has made this possible. The landscape of the city and areas of countryside is being transformed by the new technology, and the new philosophy of urban man.

In the last 50 years, road and rail transport has made it possible for people to travel longer distances to work. Following the huge urban implosion of this century there is now a tendency on the part of the relatively prosperous to move their homes back into the country. Over 1 million people went into the country-side to live between 1951 and 1966. Nevertheless the size of the cities continues to grow, and areas like the South East of Britain from Hertford to Brighton will probably become one vast urban complex in a few decades. Planning effort and public concern are needed to ensure that the land used for urban development is the least valuable for agriculture, forestry, recreation or other essential purposes.

While the total population of modern, highly industrialised communities continues to grow fast, evidence available since 1965 indicates a sharp fall in the birth-rate in Britain. The growing concern about world population, together with a desire on the part of new technological societies to live comfortably with a modest size family rather than sparingly with a large one, is taking effect. The education of women for work encourages them to return to employment sooner after marriage and childbearing, and the widespread availability of new contraceptive methods such as the 'pill' make this a practical possibility. It is too soon yet to say how far this trend will effect the overall growth of national or world population by the end of the century, but it is a hopeful sign that the new technology and its attendant psychological and social factors seem to lead to some levelling-off of population growth.

Standards of health and hygiene have improved enormously in the last century following the pioneer work of Edwin Chadwick in the 1840's. The introduction of public health regulations, machinery and public works now ensures clean water, good food and adequate drainage in town and country alike. The developments of the clothing, boot and shoe and pottery industries have played their part in improving health and cleanliness. Lighting and heating have also been vastly improved. Recent clean-air legislation in Britain (1956) and the marketing of relatively cheap primary fuels like natural gas, oil and anthracite, supplemented by huge quantities of electric power from coal and atomic energy, have all helped to provide a cleaner environment, and abundant hot water for washing and heating.

Since 1945, new synthetic fibres such as nylon have appeared from which easily washed clothes and furnishings have been manufactured. Manufacturing standards in textiles and clothing have been steadily improved. Preventive medicine has applied vaccines to large populations to eliminate dangerous diseases such as poliomyelitis, measles and tuberculosis. Standards of diet and food hygiene have risen, with new techniques of packaging and freezing or freeze-drying—ten years ago chickens were a luxury in many household diets, today they are a commonplace produced by a multi-million broiler industry. Dental decay is being slowly arrested by fluoridation of normal water-supply and by education in feeding habits and diet from an early age. Standards of health and safety are improving in manufacturing industry, in building design, ventilation, lighting and heating. All these developments heighten the need for education and human responsibility.

At the same time, it must be recorded that in Britain as in all other industrial countries, the relics from the early days of the revolution abound. It was reported in 1968 that over three-quarters of a million homes in Britain had no bath; over half a million houses were scheduled for slum clearance; and in parts of overcrowded cities, up to ten families use one lavatory. It is probably true to say that the average citizen gives less thought to the improvement of the basic conditions of life for himself and his neighbours than to the latest fashion in clothes, the fortunes of the football team, and the output of the television companies.

In the field of transport, very rapid development has occurred since 1950, and is still taking place. The majority of families own a motor car which is used for travelling to work and for leisure purposes over distances undreamt of by our forefathers. The rush-hour problem and the jammed lines of vehicles into and out of cities and seaside resorts on summer weekends constitutes a growing crisis. British vehicle registrations reached 14,804,700 in 1968, more than double the figure in 1950. By 1980 it is estimated that there will be 20 million private cars in Britain alone. A continuation of this rate of increase will extend the urban sprawl impossibly in the next decades and completely throttle the roads of these islands as the Buchanan Report suggests, yet few restraints have so far been applied other than parking restrictions in towns and higher vehicle taxes.

Throughout the world, and particularly in industrialised countries, more and more people are travelling by air and visiting other countries. With the age of the Jumbo Jet it is confidently predicted that 40 million people will be carried by them each year on long-distance scheduled flights or on holiday charter flights to other continents. Air passenger traffic is growing at the rate of 40,000 additional passengers each week carried by the world's airlines, even at 1969 fares. This enormous movement of people is a major factor in creating a new internationalism, encouraging emigration and establishing a common, technological culture throughout the world. The boom in the business of the travel agent will certainly continue to increase, barring war or slump. International man has come to Stratford-on-Avon, Salt Lake City, Stuttgart, Sienna and Sydney, with camera, passport, travellers' cheques and sunglasses. But now he is the man who lives there, as well as the foreigner with his guide book and hideous accent.

With passenger traffic growing at a rapid rate, especially on the roads and in the air, goods transport also expands ever faster. Huge oil tankers ply between the terminals of the producing countries to the jetties of refineries in the rest of the world, especially to the industrialised nations whose appetite for crude oil for petrol and chemicals grows incessantly. Container ships now cross the oceans to newly-built container ports with rapid turn-round facilities served by road and rail container systems. Bulk carriers bring millions of tons of raw materials like grain and iron ore to be processed. Air freight expands proportionately for fast or specialised deliveries, and is now becoming integrated with the container routes. Thus the primary commodities needed by modern industry, and its countless products, are conveyed faster and in larger quantities over the face of the earth, particularly between the developed countries. The monthly trade figures become a matter of interest at the breakfast-table alongside the criticism of last night's television show. Western man has become producer and economist as well as consumer and man of leisure.

Morris Ernst, the American writer, in his book **Utopia 1976** has estimated the weekly hours of work of the ordinary working citizen in the U.S.A. as follows:

DATE	*1776*	*1925*	*1933*	*1953*	**1976**
AVERAGE HOURS PER WEEK	80	60	47	40	30

The same figure for 1953 was suggested in previous speculations by Sir Thomas More in 1516, and by Bertrand and Dora Russell in 1923, but at a considerably lower standard of living in each case. Real incomes have more than doubled between 1925 and 1953, while the average working week has fallen by one-third. The position in European nations and in the U.S.S.R. is well behind this, but it is expected that the gross national product, personal consumption and average earnings in Europe will rise as fast as those in the U.S.A. in the coming decade and reach the current American level by 1980.

Ernst has also analysed the breakdown of the American worker's income in 1953/54. Most goes on primary needs such as food, clothes and housing of a high standard, and on items of technology such as new cars, 'freezers' and washing machines. Some 6% is spent on recreation, which is enough to support a huge leisure industry. $3\frac{1}{2}$ billion dollars is spent on do-it-yourself equipment; 1 billion dollars per annum on photography; over 3 billion dollars each year on radio and television and foreign travel respectively; 1 billion dollars on records, though very much less on books. Incomes in the U.S.A. are proportionately much higher than in Britain or other European countries, though the British spend a higher percentage on leisure activities, including large sums on gambling, tobacco and drink each year, and proportionately more on books.

It is estimated that the average Englishman spends some four hours each day in front of a television set. There can be no doubt that this particular mass medium is now a dominant influence in every home. Figures of viewing for historic events such as the Moon landing of Neil Armstrong and 'Buzz' Aldrin in 1969 run into hundreds of millions of people, a fact made possible by the new technology of the space programme and satellite communications.

Radio and television links bring instantaneous news to the home from most places in the world: from the moon itself it takes a little longer—about $2\frac{1}{2}$ seconds. The same telecommunications development has affected the international news coverage of newspapers and journals, though it increasingly presses the daily paper to provide considered comment and opinion rather than instant 'hot' news. A glance at financially viable newspapers confirms this trend, which in turn affects non-specialist weekly journals. The British trend follows the American pattern of larger, daily publications filled with comment and advertisements.

It has often been observed that the mass media of the non-communist world incline to fill their space with sensational, violent, criminal or sexually provocative material. This is rationalised as the portrayal of 'real life'. In fact it claims high viewing or reading ratings because it appeals to the fantasy life of the human animal, who is by nature curious, aggressive, fanciful and highly-sexed. The mass media thus provide outlets for these instincts which cannot readily be satisfied in highly ordered, modern society. The Communist mass media, on the other hand are restrained, nationalistic, virtuous and authoritarian, attempting to sustain (with decreasing success) acceptable social attitudes and a rigid 'Victorian' modesty. Other outlets for basic instincts are found in gambling—hallowed now by Government lotteries in Britain and other Western countries, alcoholism, and the experimental behaviour of modern 'dancing' or drug-taking, all of which are found in Communist countries also.

These features of the new society underline the difficulties of adjustment of modern man to his new situation. Lacking the personal ambition for self-advancement which inspired so many pioneers in the first phase of the industrial revolution, or the communal goal of nationalism which motivated the establishment of the various empires of Britain, France, Holland, Portugal, Germany, Italy, and most recently, Russia and the United States, modern man has tended to become introspective. He has channelled his drives into other outlets if he has not discovered or been able to use the new creative opportunities of leisure or technology. This maladjustment is a conspicuous feature of modern life; it has both individual and social manifestations, revealed by overcrowded mental hospitals and prisons. The perils of boredom, frustration, passivity and stagnation are the new menaces. As Professor Dennis Gabor[1] has said 'we shall need a moral equivalent of the economic game, as we shall need a moral equivalent of war'.

Fortunately, there are signs that in the team-work required for massive technical projects like the *Concorde*, and the N.A.S.A. programme, one outlet of personal and national ambition has been found. In the growing emphasis on creativity and the arts in primary and secondary education, the creative energies of the coming generation may find another rewarding outlet. Participation in cultural and political life in general, and in decision-

[1]D. Gabor **Inventing the Future** *Penguin* 1964 p.

making at work and home also appear to be necessary ways in which the instincts and energies of modern technological man can be hallowed and directed aright, but it is not clear that these outlets will be available soon enough to combat the new problems.

Modern society has been labelled with various *clichés*, each of which contains an element of truth. It is a **Consumer society**, though its productivity is greater than any previously known. It is a **Democratic society**, though increasingly vital decisions are taken outside parliament, senate or congress by industrialists, planners and bankers. It is a **Free society**, but with increasing size, complexity and planning, real choices are inevitably restricted in many areas of life. Modern society is more governed, more planned for, and more restrained by law and administration than its predecessors, even if the barriers of birth, class, education, religion or race were just as real and effective restraints in the past. The hostility to the establishment felt by many young people is thus a predictable hazard. The new society is the **Educated Society**, and education has now become both a major national investment and a life-long process of learning and retraining. At the same time, it is a society in which the new 'poor' are the less able, and intellectual merit creates new personal and social barriers. Satisfying job opportunities are open to those with necessary qualifications and skills, but not to the rest. One question which arises as a matter of urgency is: Can modern industrial society exist as a two-tier state of skilled and unskilled classes?

Religion, like many of the arts, has been pressed to fill a place in man's private, inner world in technological society. Since the rise of modern science, and the division of the inner and outer worlds of experience recommended by the French philosopher Réné Descartes, much popular religious thought has centred on personal faith and experience. It is of the greatest significance that the Evangelical revival of the eighteenth century, associated with the names of John and Charles Wesley and George Whitfield, which stressed personal religion, occurred during the first stages of the industrial revolution in Britain. This was followed by a second evangelical movement in the 1880s, led by men like Moody and Sankey. The Oxford Movement which began in 1833 with its emphasis on tradition, church and sacrament was more than a little tinged with romanticism for the Medieval past, and encouraged the idea of a separate religious world contained in a sectarian church.

Catholicism and Orthodoxy have also followed this path, especially in Britain and the United States of America. The stance of 'traditional' Christian thought and church life, bravely maintained in the face of a fast-changing society, has generally been backward-looking, individualistic or eclectic, with little to say to the religious and moral issues thrown up by the new technology and social order. Notable exceptions are the Christian Socialists and their successors, inspired by F. D. Maurice (1805–1872) and William Temple (1881–1944).

It is a sad fact that until the middle of the nineteenth century, the Churches rarely noticed the industrial revolution. Little church building occurred in the enlarged cities before 1840, and the needs of the working class were largely ignored, except by the Methodists. Happily, the Churches are now reconsidering their stance in religious thought and in Christian practice and responsibility. Both the World Council of Churches, formed in Amsterdam in 1948, and the Roman Catholic Church have been engaged in major studies of the implications of modern technology and social change. New international, ecumenical efforts are reversing the sectarian trends of 300 years. National agencies such as the Church of England Board of Social Responsibility and the Methodist Church Citizenship Department have also been working at national problems in recent years. Moves towards Church unity are now cautiously taking place. But the findings of these bodies have yet to make much impact upon the personalist, home-centred, club-like outlook of most neighbourhood churches. Meanwhile the principalities and powers of modern society continue to cut the ground from under the Church weakening its influence on the community at large. Little awareness is yet apparent of the basic issues of the time: man's use of creation; his responsibility for the future; the international and intercultural dimensions of his life; or the understanding of sin and salvation in terms of his new place in the world. The same general tendency towards an individualistic or eclectic form of traditional religion may be detected in Islam, Hinduism, Buddhism and Judaism where industrialisation has occurred.

In 1944, a few days before he was hung by the Gestapo, Dietrich Bonhoeffer wrote these words:

> *It is said that the distinctive feature of Christianity is its proclamation of the resurrection hope, and that this means the establishment of a genuine religion of salvation, in the sense of release from this world. . . . But is this really the distinctive feature of Christianity as proclaimed by the Gospels and St Paul? I am sure it is not. The difference between the Christian hope of resurrection and a mythological hope is that the Christian hope sends a man back to his life on earth in a wholly new way.*
>
> **Letters and Papers from Prison** pp. 153/4

To some of the issues posed for such a world-affirming religion by modern technology we shall return in Part II of this book.

7
Topics, Projects, Questions and Sources

1
Indicate the significance to the archaeologist of tools found on the sites of ancient settlements.

2
'The development of the city was the greatest disaster ever to overtake mankind'. Discuss.

3
'Christianity, even more than trade or conquest, has been the greatest factor in the spread of technology in the western world.' Do you agree with this claim?

4
Should the Christian Church in developing countries be concerned with providing schools, hospitals, machines and technical education, or with the preaching of the gospel?

5
Debate the motion: 'In the opinion of this house, religion has always been on the side of the *status quo*.'

6
Find out what you can about the sociological, technical and psychological factors which underly racial prejudice. How can such prejudice be overcome?

7
Prepare a fifteen minute radio script (using songs as well as the spoken word) on some aspects of conditions in mines, factories or worker's homes in the 19th century.

8
'Cleanliness is next to godliness.' Do you agree?

9
Compare the expenditure of time and money on leisure interests among members of your class or group. List the activities which give most satisfaction and try to explain reasons for this to the class.

10
If an underdeveloped country is unwilling to reorganise itself to bring about technical changes necessary for the maintenance or improvement of living standards, should it be compelled to do so by a dictator or by international pressure?

11
What are the practical consequences—positive and negative—of women undertaking full- or part-time work?

12
The Duke of Wellington is reported to have disapproved of the railway 'because it made it possible for the workers to move about unnecessarily'. Discuss the value of travel today.

13
'The mass media provide a desirable safety-valve for the dangerous instincts of the naked ape'. Discuss the place of sex and violence in these media.

14
'Religion is what a man does with his solitariness' (A. N. Whitehead). Do you agree?

Sources:

General Studies—

(Ed.) C. Singer et al. **The Oxford History of Technology** *O.U.P.* 1957.
T. I. Williams and T. K. Derry **The Shorter History of Technology** *O.U.P.* 1960.

S. L. Lilley **Men, Machines and History** *Cobbett Press* 1948.
Lewis Mumford **The City in History** *Penguin* 1969.
(Ed.) (Scientific American) **Cities** *Penguin* 1967.
V. Gordon Childe **What happened in History** *Penguin* 1968.

Technology and Society in Ancient Times—
K. P. Oakley **Man the tool-maker** *British Museum* 1965.
Lewis Mumford **Technics and Civilisation** *Harcourt, Brace* 1934.

E. W. Heaton **Everyday life in Old Testament times** *Batsford* 1948.

Technology and Society in Medieval Times—

Lynn White Jr **Medieval Technology and Social Change** *O.U.P.* 1966.

R. J. Forbes and E. J. Dijksterhuis **A History of Science and Technology** (2 vols) *Penguin* 1963.

Technology and Society in Modern Times—

R. A. Buchanan **Technology and Social Progress** *Pergamon* 1967.

L. F. Hobley **Living and Working** *O.U.P.* 1964.

Phyllis Deane **The First Industrial Revolution** *O.U.P.* 1965.

(Ed.) E. Royston Pike **Human documents of the Industrial Revolution in England** *G. Allen and Unwin* 1966.

G. Friedmann **Industrial Society** *Free Press of Glencoe* 1955.

(Ed.) Scientific American **Technology and Economic Development** *Penguin* 1965.

Morris Ernst **Utopia 1976** *Reinhart* 1955.

E. Ginsberg **Technology and Social Change** *Columbia* 1964 (especially D. Bell: **Post-industrial society**).

Technology and Religion—

C. A. Coulson **Science, Technology and the Christian** *Epworth* 1961.

(Ed.) G. Walters **Religion in a Technological Society** *Bath U.P.* 1968.

E. R. Wickham **Church and People in an Industrial City** *Lutterworth* 1957.

I. G. Barbour **Science and Religion** (Part 4) *S.C.M.* 1968.

Harvey Cox **The Secular City** *S.C.M.* 1967.

C
Technology, Culture and Change

I
Technology and Social Change

In the fascinating story of technology, certain single inventions have made a particularly important impact on the social scene. An examination of some examples will make this clear.

One decisive invention for the history of Western Europe was the stirrup. It is probable that this device was invented in China during the fifth century A.D. but it first appeared in Europe in the early part of the eighth century, becoming standard equipment for the Byzantine army of the Emperor Leo VI in 890 A.D. and for the Frankish army of Charles Martel around 825 A.D. Its use enabled horsemen to fight with longswords, lances and the art of shock combat; it is thus comparable to the introduction of the tank on the Somme in 1916. By exploiting this new pattern of warfare, which foot-soldiers were unable to resist, the Franks under Martel established an empire and a new structure of society to support it. This social organisation was the Feudal system.

The feudal class was composed of the armed horsemen, who fought and jousted in characteristic style—thanks to their stirrups. The social and economic system was so arranged that villages, land, tithes and duties centred round these knights, their horses and needs. The knights created a secular culture and patterns of chivalry, parallel to the culture of the church and interwoven with it, which lasted for a thousand years. Perhaps the most remarkable instance of the political importance of the stirrup is the Norman victory at the Battle of Hastings. The

Norman cavalry, stirruped and well-armed, outfought and outpaced the static battle-lines of the Saxon bowmen. The social and cultural consequences of that victory still remain.

Another vital single invention was the crank by means of which horizontal motion is converted to rotary motion, or *vice versa*. The earliest example again comes from China, in 31 A.D.; a pair of water-powered bellows. The Chinese also used the crank in winnowing machines from the second century A.D. The device was unknown in the West until 1420, except for the handle of grinding machines or querns, extensively used by the Greeks and Romans. In the fifteenth century it is found in the carpenter's brace, and then used for piston pumps and paddle-wheels. In 1524 it was attached to a simple spinning wheel, and then to a succession of machines leading to its first use with steam engines in 1780 by James Pickard of Birmingham. By means of the crank, power available from rotary motion—from water or wind-mills, for example—could be applied horizontally for pumping or machining, and to-and-fro motion in one plane could be applied to the wheel, making possible the revolution in transport of railways, steam-powered ships and automobiles which we take for granted today.

One effect of the crank—together with a large group of other inventions—was the stimulation of large movements of population and emigration. In addition to the drift of population to the industrial towns sited near coal-fields, Irishmen came across to Liverpool in search of work in the cotton mills or as 'navigators' constructing canals and railways. Cheap sea transport enabled thousands to emigrate to the United States to work on the railroads there. The rate of emigration from Ireland after the terrible famine of 1846 was some 200,000 each year. Iron puddlers from Swansea travelled to Germany and Sweden when cheaper and richer ores than those found in South Wales put their industry at risk. There they helped to establish the huge iron works which exist today. People whose country crafts were threatened by new machinery and methods—like the lace-makers of Warwickshire and the button-makers of Dorset—emigrated to Australia.

Similar movements took place from Germany and Eastern Europe as old trades and crafts lost ground against the powered machinery of new factories. None of these movements would have occurred without basic devices like the crank, or the human desire to maintain or improve standards of life. Here social needs and desires are clearly linked to the use of new inventions.

Proposals for the jet-propulsion of aircraft go back to 1920, and in succeeding years several inventors attempted to make a working engine. Sir Frank Whittle began work on the problem in 1929, and had a successful model on the test-bed in 1937. At that time neither civilian aircraft firms or Government military advisers gave him any support for the development of his project, although German jet engine and rocket engineers were being heavily backed by the Nazi Government. In 1939, at the outbreak of war, Whittle's idea was taken up; by the end of the war, the jet aircraft had proved its worth in combat and was steadily replacing piston-engined machines. The modern supersonic fighter and the huge intercontinental jets capable of carrying 120 vehicles each are two of the present results. Again, German rocketry underlies the U.S. and U.S.S.R. space programmes. In this instance also the demands of a nation at war were needed to implement a significant invention. It is a truism to say that the instincts of aggression and self-preservation have been the most generous patrons of human inventiveness in every age. The rate of invention always rises steeply in times of war. Once again, we see that social consciousness has to be linked to technical competence to effect a dynamic change in social life, but it is a sad comment on human nature that other instincts such as a concern for others and for beauty have been relatively unproductive of technical advance.

From this discussion of a few of the more dramatic inventions in European history, it is obvious that the relation between technique, invention and society is by no means simple. Our understanding of the history of technology leaves many questions unanswered. What is it that inspires some people to invent a new tool, or develop a new skill? Why do people come to think that luxuries should be everyday possessions? What is it that makes society reorganise itself—to the apparent disadvantage of at least some of its members—in order to use a new technique. The long succession of strikes in a number of industries about new machines and methods points to the threat which always hangs over the introduction of new tools and processes, and the threat to jobs which higher productivity seems to bring. Attempts have been made by anthropologists, sociologists, psychologists and historians to provide answers to these questions. But none of them seems able so far to supply a complete explanation of the 'psychology of technique' or the social processes associated with technical change. Simple economic and materialist answers by themselves do not explain the reluctance of a society to develop a new tool to its limit, or the unwillingness of men and women today to be satisfied with less work and more leisure for the same pay.

It is possible, however, to pick out certain factors in the history of technology which appear to be operative at moments of social and cultural change. First, there has to be sufficient number of technical elements from a previous stage available for a new step to be taken. The arrival of the internal combustion engine depended on techniques of metal production and working; it could not have occurred before the invention of the crank and precision bearing, or before suitable fuels and lubricants were available.

There appears, secondly, to be a growth of population associated with each major technical advance from the earliest times. The growth in some cases follows a technical breakthrough, as in rapid upswing of the birth-rate in the nineteenth century industrial towns; in other cases population growth precedes the innovation, thus providing either the capital resources of food, money and manpower, or the sheer economic necessity for change —or both at once, as in England in the 1740s.

Thirdly, technical progress requires a stable yet flexible economy, and this is linked to a general readiness by all sections of the community to accept change. Rigid, taboo-ridden societies are invariably hostile to technical as well as social change.

Absenteeism is widespread among Negro workers in factories of the Congo, especially among manual workers, largely because of the pressures of old social traditions from the nearby tribal centres. On the other hand, ambitious Government schemes have frequently foundered because social idealism has run far ahead of technical and economic reality. During the revolutionary period of the 'People's Republic' in Indonesia in 1948, for example, it was decreed that people should work for no more than 7 hours per day and 40 hours per week. In agriculture, the production declined to less than half the previous output because it was impossible to maintain earlier levels without new machines or longer hours of work. A society can only adapt itself economically and psychologically to a moderate state of change, or it faces disintegration.

Lastly, mention must be made of the collective will of society to change and to master a situation. This has always been a central element in technical advance. The will of community has invariably been stirred most in times of war and economic crisis. Occasionally it has been moved by persistent pressure on the social conscience, as by the Movement for the Abolition of Slavery, or the campaign for the reform of Factory Acts. The technical developments of recent decades were greatly stimulated by the two world wars, and other major conflicts like those in Korea and Vietnam. The 'Space Race' has constituted another spearhead of technical advance. Jet air travel, radar, the harnessing of nuclear energy, and computerisation are some of the results. The abolition of slavery and the restrictions on child labour encouraged the search for other forms of power in the last century. The desperate economic situation in Germany after 1945 encouraged the change which led to the famous economic miracle of the 1950s.

At the same time, technical progress has often brought disaster. The introduction of powered machines in China in the 1930s shattered the cultural tradition of the people. Dissatisfaction with the ways of his fathers in one area made Chinese workers cast doubt on the ancestral traditions as a whole and paved the way for the Communist revolution. The advent of industrialisation has brought disease, malnutrition, new strains on traditional family ties and outright mental disturbance to communities in Africa, Latin America and India. The famous UNESCO study **Cultural Patterns and Technical Change** was written by Margaret Mead and others to bring human concerns to the attention of those responsible for technical innovation in unindustrialised cultures. From the many examples provided in that study, it is clear that modern technology provides the means for rapid and far-reaching cultural changes, but at some considerable risk. The readiness of a community to use new machines and skills is related to non-technical factors in a culture, particularly to the coherence of its life, the degree of adaptability of its members, and the beliefs which give it meaning.

2

The Concept of Culture

The word culture is used in several ways. It is a word which has gathered overtones of value—a man who speaks with an Oxford accent is said to have a cultured voice, a woman who reads Beowulf for pleasure is said to be a cultured person. Matthew Arnold defined culture as the 'best that has been thought and said'. The interests of other groups in society are sometimes said to make up their 'sub-culture': a term of disdain if not of abuse.

Throughout this book, the word has been used in a different sense; that of the anthropologist who speaks of culture as the total pattern of life of a society. The behaviour, customs, taboos, rituals, institutions, educational system, dances, music, writing, art, tools and skills of a group are said to embody its culture. Thus the legends and beliefs of communities such as the near-extinct Thule Eskimos of North West Greenland disclose their culture as clearly as the ceremonies which surround the coronation of a monarch in Britain, or the drive-in movie theatres of the United States.

Although most people are not aware of the fact, their culture exercises a considerable influence over them. It shapes the way in which they see the world. It provides them with language which gives a structure to their attitudes. It offers people a picture of themselves, their society and surrounding nature. An important part of education has always been the induction of young people into cultural traditions, attitudes, values and pictures of this kind. Even the basic skills of reading, writing and number are permeated with cultural content. Thus everyone—however much an individualist he may seem—is a child of his culture. This is as true of sophisticated technological man as it is of his primitive neighbours in New Guinea, or his unlettered great-grandparents who worked upon the land.

In every society, experience and reflection—together with experiments and innovations—lead to changes. In pre-industrial societies, this change was gradual, simple developments sometimes taking centuries to emerge. The growth of writing for example was a long, slow progress. Early literary style reflects the idiom of the story-teller, as the Biblical material well illustrates. The novel—apart from epic tales and religious fiction like the **Norse Saga** or the **Book of Job**—is a feature of the modern Christian era, beginning with the age of mechanical printing. The coming of mechanisation in this field as in countless others has accelerated the process of cultural change out of all comparison with previous centuries. Parallel with this—and in large measure bound up with it—the relatively new outlook of modern science has provided a further fundamental shift of perspective for modern culture. The world of nature is seen today in a very different light from that of the early fifteenth century; it is no longer regarded as sacred or animated with spirits. Society is now viewed as a dynamic, international, controllable structure rather than a divinely ordered, hierarchical framework of social orders with fixed rights and duties in which man must work out his eternal destiny. Today the self is the subject of endless concern.

Human nature is examined from the standpoint of physiology, psychology, sociology, anthropology; it is the incessant subject of literature, drama, film and the other arts. How different this is from the accepted simplicities of earlier ages when the individual was regarded as a lord or a serf, a citizen, a national, a child of God, with clear duties and expectations.

Cultural changes in industrial society have dislocated the influence of traditionally dominant institutions like the Church, the ancient Universities and Parliament. Other centres of power have sprung up alongside them like the Trades Unions, Industrial Management, the Banks, Local Government and the educational system. Cheap and rapid transport has shifted hitherto fixed populations and broken old social structures. Social mobility has given increased opportunity to those born into classes outside the old élites. New leadership in the scientific and managerial professions has emerged to challenge the inherited leadership of the upper class. These new leaders maintain the pressure for change, and anyone with ability and ambition is able to join them if he is successful in the competition. Meanwhile, the power of the individual has been greatly increased in modern, democratic society, though this is not often appreciated.

The advent of industrialisation and the acceptance of the scientific attitude are directly responsible for this huge cultural upheaval. The rate of change is now greater than at any previous period in history. Although traditional institutions persist, their influence is reduced, and their capacity to maintain the old culture greatly weakened. Education, for long the bastion of the old culture, now stresses the need for adaptability. The undogmatic approach and a critical mind are all cultivated in the interests of the pluralist culture-situation. Meanwhile, nobody is quite sure of the direction which cultural change will take, or where he stands in such a fluid social environment. Great tolerance is shown to bizarre behaviour and experimentalism. Serious attention is now given to the cultural standards of other countries and continents. And while traditional values are tolerated by critical young people in art, music, literature, religion and morality, many are reluctant to commit themselves to any final position, to an unquestioned good or goal. A subjective, personal justification for faith, values or moral action is considered sufficient.

This dynamic, open-ended situation is a new phenomenon in human history. It brings heavy pressures to bear upon the individual, which may help to explain why so many people of all ages seek for security and some kind of identity. The question 'Who am I?' is therefore the great personal issue for twentieth century industrial man, and the question 'Where is society heading?' the unanswered social problem. Hope for the future rests on man's ability to find answers to these questions.

Religion and morality have been greatly disturbed by these changes. In place of the certainties of former generations, with their culturally-shaped religious and moral experiences, and socially-approved patterns of religious feeling and behaviour, there is now less confidence about appropriate forms of faith and morality. It has been said that secularisation erodes all religion in industrial society. If religion is to regain meaning and power, new forms appropriate to the new open-ended culture will be needed. The Christian Church in particular faces the difficult task of disentangling its teaching, its style of life and even its hymns from the cultural setting of a former age, just as it had to do in the first and thirteenth centuries A.D. This may explain the tentative tone which has crept into ecclesiastical pronouncements in recent years. The present debates about the problem of authority in the Church, the contraceptive pill, and the meaning of the word 'God' reflect the same cultural changes.

Some observers have deplored the way in which the tide of industrialisation has effected both the traditional 'high' culture of the upper classes and the 'folk' culture of the people, expressed in their language, ballads and dances, and captured for us by writers like Thomas Hardy and Peter Laslett. Professor G. H. Bantock, among others, has noted the effect of industrialisation upon this tradition as well as upon architecture and education. In his book **Culture, Industrialisation and Education** he points to the narrowing of human experience through the rigid educational system of the industrial age, and to the limiting of horizons of achievement which brought forth D. H. Lawrence's astonishing outburst:

> *The great mass of humanity should never learn to read and write—never.*

Bantock also questions the output of the new agencies of popular culture—the cinema, radio, newspapers and television, and their preoccupation with audience-ratings or readership rather than academic, aesthetic, moral or personal standards.

> *The new popular culture is a consumer's culture, in the sense that, by and large, the public passively accept what is piped to them from central distributing agencies—Fleet Street, Hollywood, the B.B.C. What is also new is the sheer volume of processed and packaged experience which is disseminated through these means, the incredible bulk of entertainment and information to which modern man is subjected. What is relevant too, is the morality implicit in these offerings, and its impersonal nature. There is no question of direct contact here.*
>
> **Culture, Industrialisation and Education** p. 23.

The debate about popular culture goes back to the writings of William Wordsworth and Matthew Arnold. Its progress has been traced by Raymond Williams in **Culture and Society 1780–1950**. It is sufficient here to note the new dimensions which have been given to the concept of culture by the rise of modern science and technology, and to raise the issue of progress in relation to technical advance.

3
The Idea of Progress

It would be difficult not to hold the view that man has progressed enormously in understanding, skill, adaptability to his physical and social environment, and mastery of it since the days of the Stone Age. Professor V. Gordon Childe wrote:

> *The progress that archaeology can confidently detect is progress in material culture, in equipment. By its improvement human societies have with increasing success adapted themselves to their various environments and later adapted their environments to their own changing needs.*
>
> **Progress and Archaeology** p. 109

Childe maintained that early science in the Near East was a practical off-shoot of material progress. Few would deny that the development of scientific understanding and philosophical reflection among the Greeks constituted intellectual progress. It will not be gainsaid either that the material and intellectual leap made from the fifteenth to the seventeenth century with the rise of modern science and technology was a genuine advance.

It is not surprising therefore that thinkers in the modern era whose concern has been the advance of knowledge and material improvement have subscribed to a general philosophy of progress. Francis Bacon (1561–1625) and Réné Descartes (1596–1650) initiated this modern attitude, breaking the grip of the great writers of classical times as the final arbiter of excellence in the seventeenth century. The doctrine of progress was developed by continental writers such as Baron le Bovier de Fontenelle and Abbé de Saint Pierre, who were followed in the eighteenth century by the **Philosophes** and **Encyclopaedists**. The idea that the course of history was moving towards some desirable goal, the idea of a future **utopia**, the idea that virtue would be rewarded and vice condemned, all these became central elements in European thought and provided the intellectual background for the French Revolution. To these hopes and ideas, J. J. Rousseau (1712–1728) added the notions of human equality and perfectibility and a passion for radical changes in society, and thus the revolutionary powder was ignited.

In England, liberals like Richard Price and Thomas Paine, and socialists such as Robert Owen compounded these explosive ideas with those of freedom and cooperative planning respectively. Moderation prevailed in England however, and sufficient changes occurred to prevent revolution. The existence of a powerful non-conformist body which allied itself with progressive social attitudes and with the hard-pressed working class ensured that the English reform movement was not anti-Christian or anti-clerical as it was in France.

The German philosopher Immanuel Kant (1724–1804) extended the field to which ideas of progress were applied to the moral order and the future of the soul. Later, Georg Frederick Hegel, the idealist philosopher, applied them to the inner meaning of the ebb and flow of history. Karl Marx developed these ideas in a new direction under the influence of the materialist philospher Ludwig Feuerbach. In his book **The Essence of Christianity**, Feuerbach reinterpreted Christianity by eliminating God and the supernatural. The error of religion was the projection of the all important human relationship of love upon the screen of eternity. Theology should be understood as anthropology, the study of man, for the notions of religion are derived from matter and should be understood in terms of human experience. Marx seized upon this negative interpretation of religion, and together with the dialectical method and philosophy of historical progress derived from Hegel, formed his famous doctrine that the conflicts of history were fundamentally economic in character, and would lead inevitably towards a future classless society in a materialistic world. To Marx must go the credit for being among the first to realise that the development of an industrial society was bringing about a drastic change in human consciousness. He appreciated that man now felt differently about the world; his relation to it was altered by his technology and its attendant way of life. Marx's concern for human brotherhood and justice in economic and political terms stand in contrast to much Christian talk about spiritual equality and supposedly God-given social differences. Marx's attitude towards the future has proved a powerful attraction for both intellectual idealists and underpriveleged workers in many parts of the world. The Old Testament undertones in his longings, for all their materialism, may be said to expose the failures of the churches of Europe to proclain a meaningful doctrine of hope at the outset of the industrial era.

The Frenchman Auguste Comte (1798–1857) built his humanist philosophy upon the notion of progress; human history being divided into successive stages—Theological, Metaphysical and Positivist. In this last stage, human society is to be governed by the laws of progress discovered by the exercise of human reason. In 1859, Charles Darwin published his version of the doctrine of progress, this time in the world of nature, in **The Origin of Species**. Ideas of evolution, development and progress have been common currency since the middle of the last century, and until two devastating world wars shattered the illusion that man was becoming a better and more reasonable creature, the assumption that man was making material, intellectual and moral progress was widespread. The Christian view that man is not perfect—or by his own efforts perfectible—was dismissed in many quarters as reactionary nonsense.

It has been held by a number of distinguished scientists like Dr Julian Huxley that evolutionary processes exhibit a movement towards a good end. In similar vein the writings of Père Teilhard de Chardin suggest that the whole creative process is leading towards a final perfection, of which Christ is the embodiment and anticipation. Criticism of these positions has come from the side of science on the grounds that such value-judgements do not belong to the realm of scientific discourse, and from the side of history and religion because they seem not to give due weight to the folly and corruption of man or the evil and cruelty in the natural order. Nevertheless, orthodox Christian faith demands a recognition of the goodness of creation, and the good purpose of God for it, as St Paul affirmed in *Romans* **8** and St John the Divine in *Revelation* **21**.

Progress has religious justification, therefore, provided it is seen in relation to the gracious activity of God evoking human response and cooperation but not if it is understood simply in terms of human achievement in technology, organisation and reason. Man's failure to achieve spiritual and moral as well as intellectual and technical progress has, according to Christian teaching, led to human arrogance, inhumanity and frustration of the kind to be seen in first century Rome or nineteenth century Britain. Today such a failure threatens to dominate us under the respectable guise of the philosophy of technique.

4
Technique—A New Philosophy?

Technique is more than machines and tools. It is the sum total of methods devised by human reason and experiment having the greatest possible efficiency in every field of human action. In its emphasis on reason and scientific testing, technique today is very different from the crafts and skills of earlier centuries. In its stress on efficiency, technique exercises a peculiar fascination as a means of gaining mastery and power in the world. In its all-embracing applications, technique comes near to being a way of life, an absorbing and satisfying mode of existence—so much so that there are those who speak of the essence of living as action, activity, 'doing things'.

This philosophy represents a new emphasis in human history, brought about by the scientific application of technology. Such an outlook poses serious problems to those concerned with ends or goals of human action rather than the action to achieve those ends. It calls in question the long tradition of reflection, contemplation and the enjoyment of ideals and things-in-themselves which has informed Christian thought and western education for so many centuries. This is not to say that there cannot be striving for perfection and value within technique itself. Every engineer, doctor, politician or skilled worker can surely be a perfectionist in his particular field. Machines need to be as efficient as we can make them. But it does mean a shift of location of such values and ideals from the future goal to the present action. The end need no longer justify the means. The means can justify themselves as well as the ends or result of any technology. This is to be seen in the case of motor-car or washing-machine production where these articles are manufactured for a strictly limited life; the phenomenon of built-in obsolescence. It may be detected in the dreams of aircraft designers for larger and faster aeroplanes, and the technical ambitions of engineers concerned with the N.A.S.A. space programme.

The philosophy of technique—'If it can be done, it must be done'—represents a major challenge to man's traditional outlook. It presents responsible people with new personal and social issues for decision, raising the urgent question of whether man was made for technique or technique made for man. In other words, the problem to be tackled is whether technique—for all its fascination and appeal to man's desire to control, make and play with things—shall be treated as a function of human life serving humanity, human freedom, and dignity, or not. In short, whether technology shall be ruled by those things which men can know only through reflection and faith as good and worthwhile, or whether man allows himself to become merely an operator of a self-justifying technical process. Franz Kafka's plays and novels of man who is a slave to the machine or the organization illustrate the fearful possibilities in a society dominated by technique. The herd-like behaviour of people in the city rush-hour or in holiday traffic jams is a warning that slavery to technique—operationalism—is not a remote prospect. As Arthur Miller showed in his play **Death of a Salesman**, there are enough people in Western society trying to sell goods which people do not really want at prices they cannot afford, to raise the issue as a very urgent one already. Slavery to technique brings apparent, short-term satisfactions to some people, but at the same time it humiliates them, taking away their identity as human beings, at best making them just like the Jones next door. It also spreads a kind of degradation throughout society, and to those unfortunates in other parts of the world who are deprived of the resources for a more responsible and less poverty-stricken way of life.

The recent history of the industrial revolution contains lessons for us if we will heed them. But the path being followed by the rich nations at the expense of the less-developed, the ever-growing rate at which basic natural resources are being consumed, and the alarming pollution of the environment by industrial societies do not give much assurance that those lessons have yet been learnt. So far Western man has behaved like the Sorcerer's Apprentice. The survival of the human race in the twenty-first century depends upon the answer we give to the challenge of technique.

5
Issues for Survival

Technology has extended man's powers in many different ways. With this extension, man has acquired new responsibilities and has been faced with decisions of greater moment to himself, his fellows and his environment. The use of a hunting spear to throw at a neighbour illustrates the beginning of this process, and the accidental release of a canister of nerve gas in the U.S.A. in 1969 indicates the kind of burden man now carries. Modern technology can be used to control birth, growth and health, and for the postponement of death by drugs, transplants or artificial environments. Bertrand Russell once said that science had put a match-box in the hands of a baby: the aggressive individual in a powerful motor-car or the unscrupulous propagandist on television proves his point.

Unfortunately, the moral teaching of the Churches has tended to stop short with the individual code of the Hebrew Decalogue or the moral law of medieval Catholicism. New technical inventions such as malaria controls, contraceptive devices or weapons of mass destruction have raised new moral problems with which these traditional codes cannot deal. The Churches have given too little attention to the establishment of what Archbishop William Temple called Middle Axioms by which the fundamental obligations of Christian life—love to God and to one's neighbour—might be applied to the new problems and new situations confronting individuals in advanced, technological society. Current discussion of 'the new morality' in Catholic and Reformed circles is concerned to discover the relation between faith and morality, love and practical choices, in a cultural situation lacking the apparently precise notions of moral law, right and wrong of former ages, but it remains primarily individualistic in scope.

The technology of groups and whole societies has forced modern man to reckon with the problem of collective as well as individual responsibility. For example, the acceptance by all members of industrial society of motor transport, involving petrol and diesel-powered vehicles, means that society as a whole must accept responsibility for the resultant pollution of the atmosphere. Similarly the achievement of high standards of diet and cleanliness, depending on the extensive use of artificial fertilisers and detergents, means that society is responsible for the risks to the soil and contamination of water and body tissues which result. More dramatically, the burden of dropping the first atomic bomb on Hiroshima in 1945 cannot simply be laid on the United States President and Air Command: other political leaders, scientists concerned with its development, and indeed the whole of the Allied fraternity were implicated in that act. It may be possible to justify the dropping of the bomb in terms of the brutal accountancy of death and destruction which is the substance of modern war, but the responsibility remains, and has led prominent scientists to speak of their particular sense of guilt, and whole communities to make acts of reparation to the Japanese.

The arrival of modern technology with its attendant democratic system thus involves everyone in the new interrelationship. Each one is involved in the group's technical commitment It is now impossible to opt out of responsibility of this kind however individualistic we might feel. Massive projects like the harnessing of nuclear power, the open-cast mining of huge areas of countryside, steeply rising consumption of fuels and metallic ores, and prestige projects like the *Concorde* or the Space Exploration programme involve the ordinary, taxpaying, consuming citizen. Each time we buy a record, drive a car or eat a bar of chocolate we are committed. This involvement presents new political responsibilities to every citizen. The siting of a new town or airport, the provision of recreational facilities, the state of cleanliness of rivers are now the public concern of all—or should be in a democratic society. At the same time the economic relationship between highly industrialised nations and pre-industrial nations—expressed in terms of interest rates on loans, trade terms and investments quotas as well as in terms of technical aid and education—is a political and moral issue too. Technological societies have not dispensed with their dependence on primary products like food, timber, oil and minerals imported from pre-industrial nations in the poorer part of the world. Each rise in standards in the former tends at the moment to widen the gap between them, and sometimes to depress the real standards of the latter. The moral issues of international relations are thus vitally involved in ordinary social choices. An urgent task for religious and humane men is the formulation of codes of international justice and group behaviour for economic as well as military issues. It will be a crucial task of international and national agencies like the International Monetary fund, the International Court, Parliaments, professional associations and the Trade

Union Congress to formulate such codes as guidelines for responsible, collective behaviour in the future.

Modern scientific technology, as we have seen, is more than a tool to be used for good or evil. It is also culturally transforming force which moulds the world in which we live and alters our relation to it. The growth of technical innovation and the changes it has brought about appear to have occurred accidentally until very recent times. But techniques and devices are now available for the direction of technological development to be more deliberately controlled, and for planned or partially planned cultural changes to be effected. So far, the majority of thinkers who have analysed the effect of technical change have spoken of the need to create 'the good life', or 'the great society' or simply permanent peace. Few strategies have been suggested for the achievement of those rather vague goals which seem to be emotive, idealistic ends with little specific content. But now these vague ideals are not enough.

One important feature of the new technological situation is the small planning group: an interdisciplinary team of experts who devote their time to give concrete reality to the future, making their proposals and projections available to industry or political leadership. The growth of Research and Development Departments in party political headquarters, major industrial firms, and elsewhere under the auspices of Government Departments like the Ministry of Technology, stems from experience in the 1939–45 war, when the teams of 'back room boys' or 'brains trusts' were responsible for radar, aircraft development and floating harbours among many other inventions. Such is the capability of today's engineers, metallurgists, chemists and physicists that planned development is a fact. Materials, tools, complex computerised design equipment are available for the production of items to precise specifications. The steady extension of this capability will make possible the technology—and the culture—of tomorrow.

With this in mind, it becomes of the utmost importance that the direction of technical development is shifted towards human, democratic, and aesthetically satisfying ends. Too much present-day technology is anti-human, demanding a slave service from its operatives; it is destructive not only in its use for military ends but also in its massive wastefulness of natural resources and its effects in polluting the environment; it is often ugly and ill-designed. If present trends continue, we may be faced with a devastated countryside covered with ill-assorted dwellings and roads choked with traffic. The land may be full of bored or neurotic people, surrounded by knick-knacks and gadgets and heaps of waste packaging, being numbed to insensitivity by incessant 'pop' music from millions of portable radios (interspersed with compulsive advertisements) and drugged by a score of pills. This can hardly be a satisfying end to the technologists dream or a model for a 'free' society.

If the direction of technological development is shaped wholly with economic, political or military power in mind, there is a strong possibility that tomorrow's world will be such a wilderness—if not an atomic ash heap. It is this frightening prospect which gives significance to some recent student disturbances. In 1968, the walls of the University of Paris were plastered with slogans: 'Death to the technocrats', 'Down with the technology of extermination', 'No to the consumer society'. The French student protest which was echoed in many other countries was partly directed against the Gaullist Government in France, and its policy of developing an independent nuclear 'force de frappe'. But it was also directed against the preoccupation of the nation with material possessions, with the ever-rising spiral of production and consumption. The American student protest movement was likewise directed not only against the war in Vietnam, but also against the 'conveyor-belt' system of higher education which is organised to serve the dominant American philosophy of material progress towards some future utopia. The 'hippie' movement is determined to find satisfaction and enjoyment in the present, not at some future date beyond retirement. This refusal to live for the future points the danger of the dogma of 'progress' in a society dominated by massive organisations. It suggests that man has lost control of his technology and needs urgently to recover a better vision of the meaning of his existence and the way in which technology should serve it.

Robert Jungk, of the Institute of Questions about the Future at the University of Vienna, has suggested that a future technology should be directed to goals which may not seem immediately profitable. He proposes such goals as cleaning the atmosphere of smoke, deadening noise, concealing factories, improving safety and health, controlling disarmament, and miniaturising products of everyday use at the same time as making them more varied and individual. Jungk gives an example of a model project on the 'Glideway System' arranged in 1966 at the Massachusetts Institute of Technology. A team of students and lecturers from different disciplines worked on the transport problems of the north-east corridor of the United States, a densely populated urban area from Boston to Washington now rapidly reaching crisis-point in road and air transport. The 'Glideway System', based upon an intensive study of available data, was hailed as an imaginative scheme of the first importance. It remains to be seen whether the immediate economic and prestigious interests of the United States will ever allow the scheme to be implemented before it is too late. Prospects for the campaign against pollution seem rather better.

The question of priorities for technology—and thus of religious and moral commitment—emerge as crucial issues for the future of society, of mankind, and of every individual alive today.[1] The possibility of influencing technological development is open to individuals who take the trouble to find out the facts and to relate their concerns and convictions to them in a responsible, disinterested manner. Although many decisions are still taken, even in western democracies, before public discussion is possible, pressure can be maintained for improving this situation; proper and effective protests can be made. Meanwhile, responsible

[1]For a discussion of the direction of technological development see R. A. Arnfeld (Ed.) **Technological Forecasting** *Edinburgh University Press* 1969, the record of a remarkable conference held in 1968.

political engagement of this kind is a clear obligation on men and women of goodwill, and particularly for those who believe that their religion provides them with certain insights into the nature and dignity of man and creation before God. This is now a major issue for human survival.

Technology has thus heightened human powers and responsibilities; its application is not only a matter of common moral concern—it is also of the deepest personal and religious significance.

6

Topics, Projects, Questions and Sources

1
'Work all you can, get all you can, give all you can' (John Wesley). Discuss this maxim for the second half of the twentieth century.

2
As a team project, trace the changes in *a*) the pattern and types of local industry, *b*) hours of work and overtime, *c*) numbers of men and women in employment, *d*) wages in your local area during the past 25 years.

3
'The age of individualism is over'. Do you agree?

4
In what ways has modern technology affected man's attitude to the world of nature?

5
Discover and compare three definitions of the word 'culture'.

6
What makes an act good or right?

7
Does modern technology encourage permissiveness?

8
Debate the motion: 'In the opinion of this house, the opponents of contraception are the opponents of progress'.

9
Read and discuss **Death of a Salesman**, Arthur Miller or **The Uses of Literacy**, Richard Hoggart.

10
Is the Good Life one of reflective thought, aesthetic enjoyment, social concern, physical activity or friendship?

11
'The journey to the moon is an unjustifiable waste of time, energy and resources'. Discuss.

12
Make a group study of traffic or community needs in an urban area of 50,000 people. Design an ideal traffic system or townplan and list the priorities together with the problems which would have to be overcome to put the plan into effect.

13
Make an exhibition of the uses to which you would like to see future technology put, with imaginative drawings, charts and plans.

14
Prepare a group performance of **The Waste Land**, T. S. Eliot.

15
'Technology is applied religion'. Discuss.

Sources:

Social Change—
Lynn White Jr **Medieval Technology & Social Change** *O.U.P.* 1968.

R. A. Buchanan **Technology and Social Progress** *Pergamon* 1965.
Jacques Ellul **Technological Society** *Cape* 1963.
W. H. G. Armytage **The Rise of the Technocrats** *Routledge* 1965.

Technology and Culture—

M. Mead (Ed.) **Cultural Patterns and Technical Change** *Mentor* 1954.
J. Meyneaud (Ed.) **Social Change and Economic Development** *U.N.E.S.C.O.* 1963.
D. L. Edwards (Ed.) **Report of Conference on Church and Society** *S.C.M.* 1968.
G. H. Bantock **Culture, Industrialisation and Education** *Routledge* 1969.
Raymond Williams **Culture and Society 1780–1950** *Penguin* 1957.

The Idea of Progress—

V. G. Childe **Progress and Archaeology** *Watts* 1944.
J. Baillie **The Belief in Progress** *O.U.P.* 1950.
D. Gabor **Inventing the Future** *Penguin* 1965.

Religion, Technology and Change—

A. J. van Melsen **Science and Technology** *Dusquesne U.P.* 1961.
I. G. Barbour (Ed.) **Science and Religion** Pt 4 *S.C.M.* 1968.
G. Walters (Ed.) **Religion in a Technological Society** *Bath* 1968.
P. Abrecht **The Churches and Rapid Social Change** *S.C.M.* 1961.

Part Two

TECHNOLOGY AND RELIGION—SOME MODERN PROBLEMS

A
Problems in Social Life

In the first part of this book, an introduction was given to the nature of technology and some of the ways in which its application is a matter of social, moral and religious concern. It was particularly stressed that modern technology, under the influence of modern science, differs in scope and speed of application from pre-scientific technology, and that it is creating a host of problems for us today which demand urgent answers for our survival. The point was also underlined that modern technology is now bringing about considerable changes in the cultural context within which these problems have to be solved. Technological man of the mid-twentieth century is a different being from any of his human ancestors.

This final part of the book focuses attention on some of the problems from the point of view of the Christian tradition, although those of other religious and philosophical standpoints—Judaism, Islam, Hinduism, Humanism—share with Christians many of the questions and concerns outlined. As in the companion volume in this series, **Religion and the Sciences**, the treatment is designed to raise the issues from the Christian point of view rather than provide answers. Further exploration of the facts is recommended in the sources listed, and consideration of the problems may be aided by the discussion topics at the end of each chapter. It is important in considering fundamental moral and religious concerns that the issues should be clearly identified and the facts appreciated before useful discussion and judgement take place. Because the problems are real ones, there may not always be simple answers.

I

Work and Production

In his book **The Wheelwright's Shop**, George Bourne gave a vivid description of the effort, involvement and satisfactions of the craftsman. Those satisfactions

> *streamed into their muscles all day long from close contact with iron, timber, clay, wind and wave, horse-strength. It tingled up in the niceties of touch, sight, scent. The very ears unawares received it, as when the plane went sighing over the wood, or the exact chisel went tapping in (under the mallet) to the hard ash with gentle sound.*

The age of the craftsman lasted effectively until the middle of the last century. Up to that time, patterns of work for many exhibited the three main characteristics of craft labour. First, the craftsman controlled operations himself. He worked in his own time to his own rhythms of effort and rest, planning and doing. Because of this he was able to live in and through his work as well as by it. He was involved in and committed to every part of the job, seeing it through from start to finish, even though his work was often hard, tedious and undervalued.

As a result of this personal involvement, craft work was individualised—the second major characteristic. It was something on which the worker could leave his mark. The silversmith, potter and mason usually did this, quite literally, but individualism was expressed in planning, design, manufacture and decoration of the finished product whether this was a garden gate, a kitchen chair, a piece of weaving—or an apple pie. There were exceptions to this in forms of work before the industrial revolution and they are important to note: mining, sailing and transport, military service, and slave labour. But in practically every other kind of work before the factory system arrived, craft individualism was a dominant feature.

The third characteristic of such work followed from this individualism. In craft activity, art and technology were united. Technical innovations like the blower for the forge or the power driven potter's wheel not only improved production, but also made for better art. Turned pots could be more beautiful than coiled pots; the gates of the churchyard made by the blacksmith after his blower had been installed, were finer than anything he had been able to make before. In the same way, the music of Bach would not have been possible before the technical development of the Baroque organ, which was itself a work of art. In the age of the craftsman, technology served human skills and art. Where craft skill is required today the same is true.

The coming of industrialisation with its factories, large machines and mass-production methods has broken this pattern of work for the majority of people. Factory work is only partly under the control of skilled workers, who do not in any case usually complete the whole product. Individuality of work is lost to the system of large-scale production and specialist activity. Products from one plant are similar or identical: one motor car is like thousands of others; one bar of chocolate like a million more.

The place of art in industry is delegated to the design section, and sometimes prostituted by the needs to sell large quantities of the product as cheaply as possible to an uncritical public. In certain sectors of industry this is not the whole story. Aircraft and bridge design are acquiring new art forms. Clothing and furniture, even when mass-produced, have high artistic quality. The pseudo-art of the Victorian period—in which useful products were decorated with 'traditional' motifs and revamped patterns to give something more than utility—have given way to a recognition of art in functional forms, and in the texture of materials well used. A V.C.10 airliner has this quality, so has a well-designed door knob and a quality plastic raincoat. Nonetheless, there is still enough shoddy industrial design and poor quality production to justify the compalint.

Many people in mass-production industries cannot regard their work as a way of life. Efforts made by companies to enlist loyalty or motivate productivity often only tinker with the basic issue. Work, for thousands of people, is a means of earning money, achieving status and power, or just being with other people. It is emphatically not a creative part of existence in its own right. A survey undertaken in 1968 showed that more than one in five women who leave manual production work in the electronics industry do so out of boredom: the cost to the industry is over £2 million each year.

As work has become mechanised, so men have become mechanised. On the production track of a motor car factory, the assemblers are themselves replaceable parts of the system. They perform repeatable acts to a relentless moving chain of units in an unnatural working rhythm. In spite of a growing degree of automation in such plants semi-skilled labour is still necessary on a large scale—and such systems are constantly bedevilled by the inevitable human reaction in the form of unofficial strikes and hostility towards the management. Successful management of such plants at present consist of a judicious mixture of financial rewards and the threat of unemployment, sprinkled with welfare provisions.

From the earlier social division of labour, which Adam Smith recommended in 1780, the industrial system later moved to the mechanical division of labour at the cost of less personal involvement and responsibility. This is to be seen in book-keeping and office systems as well as on the factory floor; and in all work where people are employed simply as operators with certain limited skills. Only the introduction of automation on a very large scale will make such work unnecessary, and at the risk of creating widespread unemployment for the unskilled and semi-skilled workers. This problem can only be overcome by changing the organisation of industry from its present emphasis on large units to cooperating smaller units requiring quite new kinds of skills in design, machine maintenance, programming and marketing. It will need a new education system as well, and new forms of employment for those made redundant.

The questions now facing western industrial society are many

—but few satisfactory answers are yet available. How far is it possible to use automation and cybernetics—control-system science—to eliminate mechanical labour? Will the minding of automatic machines be more satisfying as a form of work than being an operative? Will it be possible to bring large machines under human control directly like the engineer's lathe? Is it possible to retrain the labour force for the computer age with sufficient speed to avoid mass unemployment—however great the redundancy and welfare payments? Should the apprenticeship system be scrapped, since it ties a young person at a formative stage to learning only one trade and that in a highly inefficient manner? What will the fate of those unable to learn the skills needed for the age of automation? Will there be work for them outside the factory and office? How is it possible to overcome the widespread inertia among management, workers and investors concerning industrial reorganisation? Will smaller units of production, when fully automated, make possible human satisfactions in work? Will they lead towards the quality of product which is now possible, linked with personal achievement and responsibility for the industrial worker? How can the public be encouraged to demand higher standards of workmanship and design? Are bodies like the British Standards Institute and the Consumer Council effective enough? In what ways should education encourage people to expect quality of this kind?

The present trend in sectors of large-scale manufacturing concerns, such as the motor car and electronics industries, is towards prolonging their own existence in the present form. To do this, obsolescence is built into products such as television sets or private cars; expensive redesigning and retooling occur frequently; and sales emphasis is given to conspicuous minor modifications. Less attention is given to long life and—until recently—to safety factors. This comes about because of undue concentration on home markets and unwillingness—or inability—to reconsider the present system on all sides; shareholders, management, workers and customers. It is not yet realised that in the next stage of the industrial revolution, such conservatism will mean disaster: only those who make calculated risks, and plan for the future in an intelligent fashion can hope to meet the coming situation.

In the United States of America, over half the large companies in existence fortyfive years ago now no longer exist, having been replaced by those using newer technology introduced on the basis of predicted risk. American readiness to invest money in research and innovation exceeds by far that of other industrialised nations. This contains lessons for the rest of the world. Modern technology demands change; the limitations imposed by the earlier devices and methods of the industrial revolution must be appreciated and mastered. Responsible people need to consider their attitude to change in the nature and pattern of work, towards retraining, social mobility, and risk if we are to overcome the threats inherent in today's technology.

If it is possible—as it is now—to produce large numbers of standard products automatically, it is also possible to make them long-lasting, interchangeable and aesthetically acceptable. It should soon be possible to use automatic machines to introduce more individuality into major products in accordance with customer's specifications. Attention to these points would prolong the life of such products. This would help to deal with the appalling waste of natural resources entailed by the present mass-production system. It has been estimated that at the present rate of consumption, including the re-use of scrap, present iron-ore deposits will be exhausted in the middle of the next century. The prodigality of the mass-produced, 'throw-away' approach has to be seen in this perspective.

The extension of automation should improve the quality of the product. It may also release people from tedious manual processes for other more satisfactory work. Many more people will be needed in marketing and servicing jobs as productivity within the factory improves. With shorter hours and longer holidays, it may be possible to maintain a similar sized work-force with new automation in many industries. But this raises the issues of the status of other jobs on the selling and service side as well as within the factory, and draws attention to the recreative role of leisure.

Many industrial disputes arise over the question of status. Traditionally, employees have been paid in relation to their significance in the production process. The alteration of the process often alters the relative status of qualified tradesmen, or office-workers who are naturally jealous of their traditional position. A new process frequently brings a crisis. This kind of inter-trade and inter-Union dispute has been a major hindrance to reorganisation in the British ship-building industry in recent years for example, and is a reminder that wage-packets mean more than real cash to men and women—they represent social significance in the commercial organisation and the local community. But the arrival of automation calls in question many of the traditional differentials. Unless all employees are able to participate in the social and economic benefits of new methods, and trades unions are whole-heartedly cooperative and flexible towards retraining and new demarcations of work, it is unlikely that the necessary changes required to escape from old, wasteful, inefficient methods will be rapidly achieved. The amalgamation of trades unions; the improvements of working conditions at shop-floor as well as office-staff level; the elimination of class distinctions between wage earners and salaried staff; better management supervision and communications—all these changes in commerce and industry are necessary in the age of automation if work is to be a matter of pride and responsible service within the community.

At the same time, new, automatic processes are improving earnings and allowing more leisure. The use of money and leisure time is now a matter of great importance in society. New opportunities are appearing for creative and rewarding activities outside a man's primary work. The idea of a 'second profession' is now a possibility so that in addition to a basic job, a second skill may be exercised in social service, education, the arts or the ministry of the Church. The 'Do-It-Yourself' craze is a promising start in this direction, but much more is possible if the facilities, education and encouragement are given.

In the Christian concern for work and production, the basic questions to be asked are about the nature and dignity of work, its implied relationship of man to his neighbour, and the world from which resources are taken and to which the products of work go. A vast proportion of human life is spent—and will continue to be spent—at work. It must be a major task of Christians to see that alongside legitimate profits and wages attention is given to the conditions, relationships and proper satisfactions of work, and to the responsible use of material and skills. Similarly, Christian concern for leisure should require that it shall be creative, worthwhile and enjoyable, enabling people to renew their strength, vision and relationships with man and God. Christian judgement and influence must be informed by the convictions that 'the earth is the Lord's' and man's fundamental role is that of service and worship.

The Christian concern for work is not simply a romantic yearning for the old ways after the manner of William Morris and the Pre-Raphaelites. It stems from the conviction that God is Creator continuously at work, that man is a fellow-worker with God, and is set in the world as a steward of many talents to tend and care for creation. Throughout the Old Testament work, conditions of employment and rewards are taken with the utmost seriousness. The skilled worker is said to 'show the Spirit of God himself. Work is the description Jesus gives to his ministry, for which he prepared in the carpenter's shop as well as in reading and reflection. Work is necessary to achieve human dignity, to provide due service to one's neighbours, and to live responsibly. The person who refuses to work cannot express his true nature or fulfil his social obligations. Christians believe that the objectives of work need constant scrutiny since the worship of the work of men's hands is blasphemy, and motives for work, like everything else, can be poisoned by false pride, greed and laziness. Thus Christian concern for work begins with the doctrines of man and creation, and ends with the doctrines of redemption and sanctification.

2
Organisation and Community

One of the growing points of industrial practice in the second half of the twentieth century is the participation of workers, office staff and managers in consultation about the organisation and aims of the enterprise. Already a measure of consultation goes on between management, unions and workers' representatives about certain aspects of industrial organisation in some firms. Some companies, like Imperial Chemical Industries, operate profit-sharing schemes and offer shares to employees—thereby giving them a voice in company policy and a stake in its success. The criterion of a good firm is no longer one which offers generous bonuses and extensive recreational facilities. It is one in which the organisation is something of a community enterprise, allowing everyone to contribute to fundamental decisions, especially about the structure of the system. It has been found that participation of this kind is more potent in encouraging men and women to give of their best than the imposition of the new methods on the basis of 'time-and-motion' study alone. Without such involvement it has been estimated that people are functioning at only 60/70% of their capacity.

This approach to industrial organisation marks a significant change from the traditional idea of industry: the view that workers hire their skills, energy and time to the bosses for as much money as they can get, and the bosses exploit this labour by rigid discipline and tough oversight. This kind of change follows the transition from owner-managers of factories to paid managers, and the influence of trades unions and education on the workers' situation. It represents a new attempt to make industrial organisation serve human as well as economic ends, and takes account of individual responsibility in a democratic society. But so far it is not in general use.

Similar moves towards participation—sometimes described as democratisation—are being made in universities and colleges of higher education; a fact which is certain to influence other organisations in the course of time.

The growth of cities with large and complex departments of public administration has created serious problems for community life. Cities such as London or Birmingham stand at the middle of huge conurbations of many millions of people. The sense of community and local identity which marked village and town life in earlier centuries has largely disappeared. This has been aggravated by the centralisation of city administration, largely on economic grounds, and by the high density housing projects which have multiplied since 1945. Local community life has been impoverished; impersonal blocks of flats lacking community facilities have reduced the possibility of simple, friendly meeting; the need for workers to travel long distances from home to factory, shop and office has added to the breakdown of the community situation. These are all the end-product of inadequate planning on the basis of the old technology—coupled with a reluctance on the part of ratepayers to contribute what is necessary for a community-based administration in a large urban complex.

One significant result of this is the poor voting figures in local elections. A 30% turnout is reckoned to be a satisfactory poll by local politicians, but in fact it represents a serious lack of involvement by the majority of citizens in community responsibility. Fundamentally this is a structural problem which the application of the newer technology should be able to change. The participation of every citizen in debate and decision-making for his community is the aim of the democratic system, but this can only be realised if natural community life is made possible. Responsibly planning for community development must guide the application of technology in building, transport systems, town and regional development in the future. It is vital that lessons are learnt from the mistakes made in recent decades, when village and suburban communities were suffocated by the bureaucracy and short-term thinking of city developers.

The principle here is that organisations and administration

should serve community life rather than master it by their own criteria of convenience and efficiency. Every organisation tends to justify itself on those terms, and to reduce people to predictable statistics. The dangers of this in the United States have been well described by William H. Whyte in his book about **The Organisation Man**, and much of what he writes is applicable in Britain too. The values of this type of person, Whyte suggests, set the tone of a society. Such people are the mind and soul of the great self-perpetuating corporations, government departments, and even the churches, the dominant members of industrial society. They run the organisation, not at the top, but in the second levels of management. Their virtue is loyalty to the system, come what may. Their conformist outlook makes them adaptable, pliable, and interchangeable as they move about from one suburbia to another. They are the product of the organised society, which is itself the product of technical change.

Such people do not contribute much to the **direction** of society, or to creative community life, inside or outside their organisations. De Tocqueville once wrote that if America ever destroyed its genius, it would be by stressing what he called the social virtues at the expense of others; by allowing the individual to become the slave of popular opinion or the prevailing system. His prophecy is one which applies to all organised, industrial societies which cater for the consumer, the convenient statistical unit, or the ideologies of private enterprise or social dogma at the expense of the responsible community life.

Inevitably, a complex organisation needs planning and authority; equally inevitably, it simplifies the human situation to terms it can manage, like the symbols of a punched card, and tends therefore to treat people as less than human. As a result, such an organisation can produce material prosperity, but it can just as easily destroy human community, and dehumanise the individuals associated with it. Does the organisation need to do this? Can the system, in spite of its rational methods, be made to serve people? Must the technological society obliterate the individual's freedom or his power to discriminate and differ?

Concern has recently been expressed about the introduction of computer information-storage in banking houses and government departments, and the use which might be made of confidential information about people. Is it possible to legislate effectively to protect the rights of the individual in this matter? Should the selection of people for jobs be done by mathematical devices—or by other people with whom they will have to work? Is it possible to protect the interests of the individual against the steam-roller of government? Is an Ombudsman enough? How can the planners be required to take their decisions after satisfactory **public** discussion, and not in the back rooms of the corridors of power? Current debates about regional government, regional planning and the reform of Parliament touch these issues and must be of vital concern to Christians and others concerned for human dignity and community life. The recent wave of unrest in colleges and universities is concerned with the same questions.

The problems of the organisations of large groups in a truly democratic society are forced upon us by the advance of technology, and the increasing application of scientific categories to social development. But how can the individual rightly fight the threat of suffocating organisation, and to what precise ends? Modern technology raises the problems of participation, protest and social purpose in an acute form.

3

Living in One World

In 1849, the ship **Sea Witch** sailed round the world in 194 days: in July 1968, a Vulcan jet aircraft circumnavigated the earth in 21 hours. In 1850, the first steam ships crossed the Atlantic in under four weeks: in 1972 **Concorde** will do the journey in under four hours. Radio and television communications link continents now as effectively as if people on the other side of the world were living next door. News is normally international—local news seems quaint and trivial.

Technology has created one world, making everyone's affairs our own; in Czechoslovakia, Nigeria, Mexico or Vietnam. Technology is helping to form a common culture. The same 'pop' music can be heard on Radio Hong Kong as on Radio Luxemburg or B.B.C. Radio One—with a slight variation in vocal noises as a concession to local languages. The same style of clothes is to be seen in California, Rome, Tel-a-viv, Cape Town or Montivideo. The same detergents, cigarettes, biro pens, spectacles, motor cars or films may be seen in Tokyo, New York, Paris, Nairobi and Brisbane. Orchestras, theatre companies, comedians, dancers travel all over the world—with a few exceptions—sharing cultural forms, values and achievements. The new culture is an international reality.

The great cities of the world are increasingly international places, throwing up settlements of different racial, cultural and religious groups. The concept of nationhood has been modified by the technical revolution towards the notion of the pluralist, international, interdependent society, consisting of many groups living together, holding different convictions, attitudes and cultural backgrounds in tension.

The issue of race has only arisen because technology enabled Europeans to dominate other racial groups through the first spread of trade and conquest during the fifteenth and sixteenth centuries. This success persuaded the white man that he was a superior creature, created to rule and dominate his brown or yellow fellow-beings. The further development of western technology created a new kind of dependence of coloured peoples on Europe and North America resting on economic relationships and known as neo-colonialism. Now the internationalising of the big cities is bringing the legacy of white arrogance home, stimulating fear, prejudice and widespread intolerance. The tragic need for the 1968 Immigration Act in Great Britain underlines the racial problem created by our use of technology down the centuries. Sociological investigations in the United States have shown that the race riots in American cities have been caused primarily by the frustrated, educated negroes who are denied opportunities of employment and advancement available to whites. (See *Immigration and Integration* by Clifford Hill, Pergamon).

The fact of living in one, technological world thus raises major problems: Must modern technology produce one culture throughout the world? Is this a good thing? What should our attitude be towards other cultures—such as the Welsh or Irish or West Indian—preserved in parts of our own society or in the way of life of people living among us? In what ways should technology be used to overcome barriers between nations and cultures? How can science and technology help to combat the problems of racial prejudice and conflict? In the present international situation, how can technology be shared without prejudice to deal more effectively with earthquakes or other natural disasters, with besetting problems of famine, ignorance and squalor? Are the existing agencies of the United Nations adequate to this task? Should the ability to grasp the implications of technology for the world rate more highly in our idea of a national leader than that of the traditional party representative? Can technology help men of different languages and cultures to understand each other, rather than enable them merely to make more frequent diplomatic noises at conference after conference? How can we ensure that modern technology continues to bring people together, and is not used to divide or destroy?

Fundamental to many of these questions is the relation of faith to culture. It is assumed by many that the inherited culture of Britain, Europe or the United States is the only expression of Christianity possible. A change in that culture represents a necessary abandonment of Christian faith. Leaders in South Africa and Rhodesia have frequently spoken of their desire to preserve 'Christian' standards and 'Christian' culture and have used this as a justification for the policy of white domination, with black and coloured 'separate development'. But the identification of Christianity with the human culture of any period confuses the kingdoms of this world with the Kingdom of God. The lessons of Christian history indicate that Christianity has been expressed in several cultural contexts—Jewish, Roman, Greek, Medieval, Renaissance, European, American, Indian and Chinese. The temptation to confuse Christianity with a particular culture embodied in certain customs, standards, ways of thought and assumptions is a strong one, especially when that culture is threatened. But Christian faith rests in a living God who is at work in history—not in an absolute code of behaviour or a traditional set of dogmas or cultural concepts. The disentanglement of religion from passing cultural forms is a difficult task, demanding faith, hope and a good deal of charity. With the growth of an international, pluralist cultural scene in the relatively short space of twenty-five years, such disentanglement is a matter of the utmost urgency lest our convictions turn to prejudice, and our principles to petty partisanship. This is what the Ecumenical Movement and the so-called 'New Reformation' in the Churches is about. It must also affect the nature of religious education in church and school.

4
Media of Communication

When Gutenberg discovered—or rediscovered—printing with movable type in 1454, he made possible one of the greatest advances in human social and cultural life. The printed word stimulated reading, thinking, and art among an ever widening circle of people. The arrival of quantities of literature in Africa today—educational, political, religious, pornographic—is creating a colossal demand for literacy and western kinds of education. It was no accident that the scientific revolution occurred as printing made possible the spread of ideas, diagrams and recordings. It was no accident either that the rediscovery of man—his thoughts, feelings and aspirations—in the Renaissance occurred when the printing press began to make available literature, poetry and art to an increasing public. The novel is virtually unknown before the sixteenth century; poetry is mostly a private affair before the seventeenth. Printing made public and popular the art of writing, introducing literate western man to beautiful and powerful ways of using words. Printing altered the rate at which ideas and information spread.

Today book-publishing and libraries flourish: over 17,000 novels are published each year; learned journals spawn and multiply incessantly—the total number of scientic papers published each year is now well over a million, and rising rapidly. The threat of this escalation is making the provision of new electronic information storage and retrieval systems a matter of great urgency in many fields of research and development. Early in the twentieth century, the popular press—led by Lord Northcliffe's **Daily Mail**—was the first medium to create and exploit a mass market. Northcliffe's paper used all the modern gimmicks to attract and excite its readers; it was, as one telling phase put it, 'the new irresponsibility in terms of the education of the new democracy'. The press, like other mass media, is faced with the dilemma of printing what its readers want in ways they relish, or printing a more staid and far less attractive version of the same story. It is a sad comment on public as well as press standards that sensationalism is often allowed to distort the truth. A further problem arises frequently when the actions of an individual are thought fit for public ridicule 'in the public interest'. The price of a free press has included the cruel destruction of people as well as the exposure of matters of genuine concern.

If the mass newspaper paved the way for other mass media, it looks now as though it is being superseded as a means of conveying news by radio and television. The cinema pioneered this new kind of appeal, using—like television—the senses of sight and sound to communicate to its audience.

As with the mass newspaper, the other mass media raise

religious and moral issues of great importance. What is to be communicated; truth or falsehood, biased or unbiased fact, considered or ill-considered opinion? Are the mass media to be used to pass on agreeable opinions, facts palatable to the government, for example, or should they have independence, responsibility and freedom to criticise? Should their criticism be kept within any limits—and what is to prevent it being destructive? Failure to resolve these problems has led to dictatorship and barbarism in many countries. Hitler used the mass media for propaganda purposes to terrible effect. The struggle for freedom of the media was the focal point of the abortive Czechoslovak reform movement in 1968.

The mass media undoubtedly exercise a very considerable influence upon people's opinions, attitudes and tastes. Their very existence raises important questions. For example, what standards should apply to the content supplied by the media? What degree of violence should be shown on film or in a television programme? How far should news media invade personal misery? What are the limits of an interview? What measure of sexual exposure is appropriate in broadcast pictures? How far should ridicule and satire be taken? It is felt by many that the daily diet of violence, horror and inhumanity recorded in the newspapers or presented on the television screen have the effect of numbing sensitivity to suffering. On the other hand, some consider that the comfortable westerner needs to be shocked from his cosy indifference to the brutal facts of life, and must meet the world from which his prosperity and self concern have isolated him. Precise evidence of the effect of the mass media on political judgement, private feelings or social responsibiity is notoriously hard to come by. What is clear is that the media **do** have an effect, the nature of which may have much to do with the disposition of the viewer, listener or reader.

Not only is the content of the output of the media a matter of concern, but its manner of presentation also. No final separation can be made between content and treatment in an objective judgement of media material. Does it matter, for example, if religious films are bad films so long as they are 'religious' enough in content or intention?

The question of standards raises the issue of control. Who is to say what is suitable material for a newspaper article or television film? Who is to determine what is a proper way of treating the material? Who can say what is artistically satisfactory? Is any form of censorship needed to prevent corruption of a susceptible proportion of the community exposed to the media—remembering that bad art is probably more corrupting than bad language? Are the public sensibilities of editors, and producers, coupled with the laws of obscenity, libel and slander a sufficient safeguard for the public? Should the Press Council be responsible in any way to Parliament as the Governors of the B.B.C. and I.T.A. are? As we lack formal censorship except over films, how can people be educated to discriminate between good and bad literature, radio, film and television? In an area so exposed to emotional reactions, are we to say that sincerity, interest, enjoyment and curiosity are sufficient to justify practically anything? In other words, is there any yardstick for popular culture?

In the present situation in the western world, the responsibility of choice and discrimination in these matters rest largely on the individual. Fears have been expressed that the policies of those who run the mass media and who are naturally concerned for readership numbers or viewing ratings are helping to create an amoral, mediocre cultural environment, with a strong emphasis on entertainment and the titillation of emotions. Others say that the use of the media, particularly by advertising interests, is creating a materialistic society. In reply, it is argued that the media are slowly helping to create a mass culture of depth and quality, and this is not something which can occur overnight. The problem of opening the eyes and ears of one nation to the whole reality of the world's life, and at the same time of upholding the highest standards of excellence achieved in the high cultures of mankind is indeed a difficult one. But it is one which cannot be side-stepped if we are concerned with the quality of life in the technological era, and it presents a challenge to the educational system.

Technology is having an increasing impact in education. Recent innovations range from teaching machines, language laboratories, overhead projectors and closed circuit television to schools broadcasting and the Open University. The skilful use of technical aids has unquestionably made learning by young and old more palatable and effective. But to those who have proposed the wholesale and exclusive adoption of teaching by technology, the answer of educationists has been that the educative process is essentially an interaction between pupil and teacher in the context of a community of learning. The question posed here is what is education for—and of what, therefore, must the educational process consist?

5
Weapons of War

When the first atomic bomb was dropped on Hiroshima in 1945, a cold shudder ran down the spine of every civilised man and woman. The ghastly after-effects of atomic radiation from the nuclear explosion reinforced the feeling of sickness and revulsion at this latest weapon in the armoury of scientific man. It has not contributed to the peace of mind of Western man that it was his bomb, and that he was the first to drop it. The fear of nuclear war has remained at the back of his mind since then; it has shaped international politics, particularly between the U.S.S.R. and the U.S.A.; it has introduced the concepts of 'ultimate deterrent', 'nuclear capability', 'megadeaths', and 'the politics of terror'. But such is the nature of modern scientific technology that there is no way of going back to the era before the atomic bomb. All that can happen now is an increase in number of countries who possess it; an acceleration in the production of complex defence systems; and an attempt to achieve the effective international control of nuclear explosions for whatever use. All these developments have begun to take place in the last few years. A tremendous effort is needed to contain and master the problem for the sake of mankind. Meanwhile, the unease remains and huge resources are expended.

While the size and long-term effect of nuclear explosions distinguish atomic weapons from more conventional military horrors, the issues raised by them do not differ fundamentally from those presented by any military tool. Weapons are means of destroying the enemy, or of frightening him into submission. They can be used in self-defence or in aggression. They can have immediate or long-term effects on human life; not only on the active participants, but upon families, societies, and even the unborn as well. The aftermath of war was known by primitive man as well as by shell-shocked soldiers of the 1914–18 war, or men, women and children demented or maimed in Korea or Vietnam. It is in the nature of military technology to do these things.

Modern warfare is said to differ from that of the Middle Ages in its indiscriminate barbarity. Certainly it is conducted on a larger scale, and little attempt is made to distinguish between combatant and civilian in the heat of battle. It is more total, more varied in style and less personal in execution. But the gruesome details of medieval battles when men cut one another to pieces over the bodies of their fellows is hardly to be preferred to rockets, bullets and even napalm. What happened to women and children then was little better in the end. At least the organisation and conduct of war leaves room for individual skill, judgement and occasional humanity today, instead of blind obedience to stupid generalship immortalised by the charge of the Light Brigade. For such crumbs of progress we may be thankful.

Military technology of all kinds, from pistols to intercontinental ballistic missiles and from brainwashing devices to germ warfare, raises the moral and theological issues of power. There is no escaping the problem of this kind of power in a world of many nations, human aggression and advancing technology.

From the moment a primitive man living in a family group raised a hand axe against a rival group, the problem existed. Protests against weapons of mass destruction serve as a useful reminder to the human conscience that man has such knowledge and the propensity to use it; but protests do not usually face the real problem—the problem of the right use of technical power. What has religion to say about this?

First, religions differ in their attitude to life in this world and the next, and thus over the basic problems of life we face. Buddhism for example, which encourages utter detachment of the soul and isolates religion from life, has little to say except that violence is a bad thing, and none should take any form of life. Christianity, however, being concerned with God's kingdom in this world as well as in heaven, has much to say about human nature and responsibility. Christianity is therefore concerned with the use of all kinds of power, for war as well as for peace. The language of Christian religion—salvation, redemption, faith, hope, love—has a direct reference to the facts of human life on the collective, social level as well as for each individual. The biblical idea of salvation is a social concept: it concerns the community of all God's people. The New Testament idea of love is a practical, social activity as much as an attitude towards individuals. Jesus was involved in a complicated political situation, and used highly charged political terms like the Kingdom of God. He did not simply go about curing souls; he handled **power** and spoke much about its use.

For some Christians, the New Testament implies a pacifist approach, that is to say the renunciation of all forms of violence in any situation whatever. This approach was adopted by Christians in the first few centuries before the Church became established as the official religious body in the Empire. Pacifism raises acute problems, however, for the social conscience. If one is unprepared to use force under any circumstances, even for the maintenance of law and order, or the protection of those wrongfully attacked, how can one responsibly belong to, or receive the protection of, society? Most pacifists adopt a compromise position rejecting the use of force in war or for major political ends. They defend their position by asserting what they are witnessing to the best interests of society by such refusal and are in fact defending society by rejecting violence. The case is expanded by the late Professor Charles Raven in **The Theological Basis of Christian Pacifism**.

For others, pacifism is not the best way in a sinful world. They maintain that there is little evidence that pacifism works, but rather it invites aggression. The social responsibility of the citizen includes the bearing of arms, and the defence of society, its laws, land and members.

Christians who believe this hold that the realities of the social, political and technological situation have to be grasped **as a matter of faith**. Human choices and human actions, they maintain, are always imperfect and sinful. Motives are always mixed, and no one path can be absolutely right since every action involves some kind of suffering. All human life is therefore lived under the shadow of the cross of Christ. But just because of this we have to make the best decisions we can in the situation and seek forgiveness: the resurrection of the crucified Christ means that we are accepted and forgiven nonetheless. This view leads to active Christian involvement in the hard facts and unpleasant decisions of social, political, military and technological life. As Martin Luther put it, Christians have to 'sin boldly' rather than seek to preserve their souls. This view requires Christians to discover the facts, make realistic appraisals of any problem, and take their share of responsibility before man and God for the decisions they take.

This kind of thinking about politics and the weapons of war is carried out by the Churches Commission on International Affairs, an associated body of the World Council of Churches. A sample of its approach is to be found in Alan Booth's **Christians and Power Politics**, which makes a clear statement of the position and its application to the issue of the nuclear deterrent.

Modern weapons compel people to make up their minds urgently about the use of this terrifying kind of power, and the bearing of their beliefs upon it. In practice, the sheer enormity of these devices and the dangers involved in their proliferation have forced many realistic thinkers to take the view that it would be better to suffer invasion and oppression than to engage in nuclear combat. But the fact that we have such weapons in a state of readiness may mean that such an option may not be open to us. A deterrent ceases to have any effect in the bringing about of a stable situation if one is not prepared to use it as a last resort. The hope is that the deterrent will deter, and time may make for better understanding between the great powers. If we can play for time long enough, there are signs that technological change and peaceful coexistence may bend rival ideologies and reshape the world into a new unity. That is the only hope, humanly speaking, for getting rid of such weapons for ever.

Equally obscene items in the armoury are biological weapons of germ warfare. The consequences of using such things—for the whole environment as well as mankind—are completely unknown and quite terrifying. One gramme of the bacterium *Clostriodum Botulinum* could kill thousands of animals and people. Slightly less alarming are chemical weapons with the power to put opponents under sedation for a limited period. The development of this kind of weapon is likely to proceed rapidly in future, making possible 'the bloodless war'. The implications for morality remain the same, however: under what circumstances, if any, is one group or nation justified in using force against another? Is a form of pacifism required by religious conviction or sheer expediency in the present situation? How is love related to law, and its enforcement? Is there anything worse than taking another life? If force can be justified at all, does it matter what methods are used? Are the hydrogen bomb or the bacterial spray no more than the extension of the hand-axe from a moral point of view? Are some weapons less immoral than others? Is it ever possible to go to war in the name of God?

6

Topics, Projects, Questions and Sources

1
Why work?

2
'In the modern world, we must regard our real life as our leisure, not our work.' Discuss.

3
Arrange a visit to a store or factory. Prepare a group report giving an account of the different kinds of work you see, the reasons given for doing it by the employees, and the satisfactions they find in their jobs. Discuss your findings.

4
Prepare a five minute talk evaluating a common product (e.g. soap powder, record, item of clothing). Indicate the cost, the manufacturer's claims for the article, workmanship, design and effectiveness. How does your evaluation compare with that in (say) **Which?**?

5
Debate the motion: 'Strikes are immoral and should be made illegal'.

6
As a class project, dramatise three short scenes to illustrate three factors needed for the smooth and effective organisation of a group of people.

7
Read and discuss **The Insect Play** by Karyl Capek **or** **Saturday Night and Sunday Morning** by Alan Sillitoe.

8
Prepare a chart of the political structures, civic organisations and public services of a city, county or county borough.

9
Is it right for people of different races or cultures to marry?

10
Write a 250 word criticism of a television programme.

11
Debate the motion: 'In the opinion of this house, all Christians should be pacifists'.

Sources:

General—
E. R. Wickham **Church and People in an Industrial City** *Lutterworth* 1958.
R. Taylor **Christians in Industrial Society** *S.C.M.* 1961.
E. R. Wickham **Encounter with Modern Society** *Lutterworth* 1964.
(Ed) I. G. Barbour **Science and Religion** (Part Four) *S.C.M.* 1968.
Report of the Uppsala Assembly of the World Council of Churches *W.C.C.* 1968.

Work—
George Bourne (Sturt) **The Wheelwright's Shop** *C.U.P.* 1923.
E. P. Thompson **The Making of the English Working Class** *Penguin* 1969.
G. H. Bantock **Culture, Industrialisation and Education** *Routledge* 1969.
A. Richardson **The Christian Doctrine of Work** *S.C.M.* 1955.

Organisations—
W. H. Whyte **The Organisation Man** *Penguin* 1954.
J. A. C. Brown **The Social Psychology of Industry** *Penguin* 1958.
H. Cox **The Secular City** *S.C.M.* 1965.

R.T.—K

One World—

(Ed.) D. L. Edwards **Christians in the New World** *S.C.M.* 1966.

(Ed) J. B. Bennett **Social Ethics in a Changing World** *S.C.M.* 1966.

(Ed) E. de Vries **Man in Community** *S.C.M.* 1966.

(Ed) E. Barker **The Responsible Church** *S.P.C.K.* 1966.

S. Rose et al **Colour and Citizenship** *O.U.P.* 1969.

Mass Media—

R. Williams **Culture and Society 1780–1050** *Penguin* 1963.

R. Hoggart **The Uses of Literacy** *Penguin* 1958.

Pilkington **Report of the Committee on Broadcasting 1960** *H.M.S.O.* 1962.

Warfare—

A. Booth **Christians and Power Politics** *S.C.M.* 1961.

C. E. Raven **The Theological Basis of Christian Pacifism** *F.O.R.* 1952.

(Ed) D. M. MacKinnon **God, Sex and War** *Fontana* 1963.

B

Controlling Man and the Environment

I

Population

It is notoriously difficult to estimate populations before the nineteenth century. Deductions have been attempted from the area of ancient cities, tax returns, trade accounts and the statements of early historians of the period in question. It is generally believed that over the whole world, the population increased slowly at rates between 0.1 and 1.0 per cent per year until it had reached some 350 million by the end of the sixteenth century.

The first censuses were taken early in the nineteenth century in Britain and Sweden—in spite of considerable opposition—and it became clear that in both countries a substantial increase was occurring. The figures for England, Wales and Scotland over the next 150 years are revealing:

Population of England, Wales and Scotland (in millions)	*1801*	*1851*	*1901*	*1961*
	10½	21	37	51

In more recent times, it has been possible to obtain reliable population statistics for countries throughout the world. United Nations demographers have given these figures for the growth of the world's population since 1920:

World population (in millions)	*1920*	*1930*	*1940*	*1950*	*1960*
	1,810	2,013	2,246	2,485	2,990

Estimates made in 1963 based on a medium projection—that is a somewhat modified rate of increase compared with present

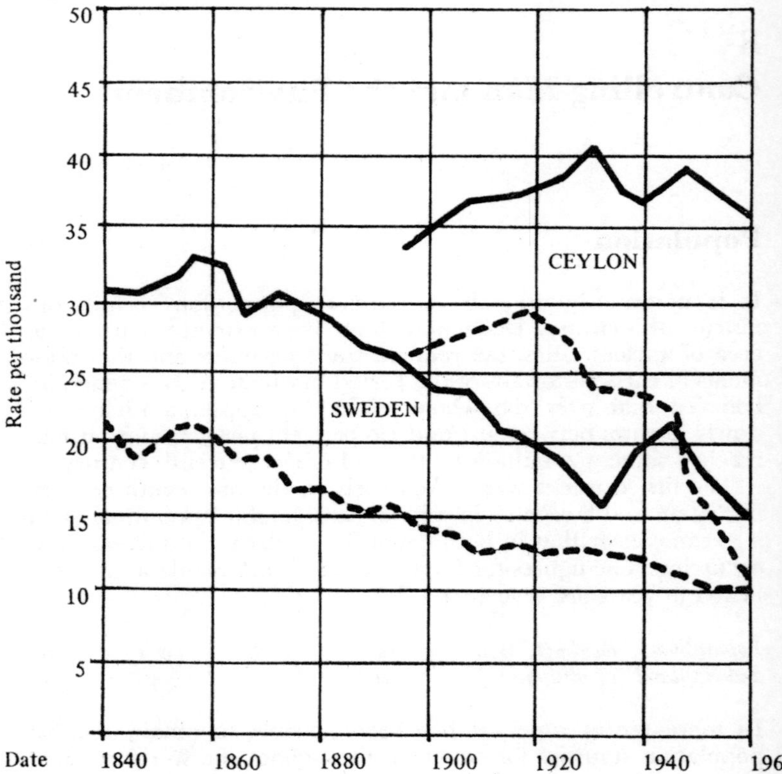

figure 2

Demographic Pattern of Industrialised and Non-industrialised Societies (Sweden and Ceylon) **shown by changes in birth rate** (solid line) **and death rate** (broken line). (After Kingsley Davis *Population* in **Technology and Economic Development** *Pelican* 1965.)

growth-rates—indicate that by 1970 the world population will be 3,574 million, and by A.D. 2000 it will be 5,965 million persons. It should be noted that if current trends continue unchecked, these figures will be 3,626 millions and 7,410 millions respectively. In other words, the population of the world is likely to double by the end of the century, unless some dramatic change occurs. It has been estimated that if this increase continues there would be as many people living on each acre of land in 250 years time as now live in Greater London. Another calculation has predicted that by the year A.D. 2400, there will only be one square yard of land for each individual. These predictions are constantly being revised in an upward direction as rates of population increase rise. Quite apart from the terrifying overcrowding in prospect, it has to be remembered that each person consumes per year the food produced by at least half an acre of land—farmed by the best modern methods. Thus starvation, conflict and disaster stare us in the face from the near future unless drastic changes can be made, yet the urgency of the problem is not fully recognised by the general public.

The complex interrelation between population increase and industrialisation has already been mentioned (Chapter 3). What is now clear is that two main factors are at work in bringing the unprecedented increase of numbers about.

During the first 75 years of the nineteenth century, the birth rate in England and Wales remained at 35 per thousand per year, a relatively high level. By 1938, it had fallen to about 15 per thousand. Spurts in this rate occurred after the second world war (18 per thousand) and again in 1962. But more recent figures have shown a sharp fall since 1965 indicating that the limitation of births is increasingly in the interest of modern living. The pattern in the United States also showed a high birth rate in the earlier phase of industrialisation. An expected fall in the 1930s did not materialise, and from 1940 to 1965 the rate remained well above 20 per thousand.

During the last 30 years, the death rate in the United Kingdom and the United States has continued to decrease slowly to 17 per thousand and (approximately) 10 per thousand respectively, giving a life-expectancy of well over 70 years from birth. This has accounted for the continuing increase of population in both countries—especially in the United States, which is expected to have a population of some 4000 million by the year A.D. 2000.

The other factor, predominant in less developed countries, is the **rapid** fall in the death rate compared with a constant level or slow decline in the birth rate. Between 1948 and 1965, the birth rate in Mauritius fell steadily from 41.5 to 35.5 per thousand, but the population of the island grew from 429,000 to 741,000; the rate of increase going up from 1.4 to 2.7 per cent. This was largely due to the control of malaria. The population of India grew at the rate of 5 million per year between 1950 and 1955; between 1955 and 1960 it grew at the rate of 7 million per year, and current estimates put the present rate of increase at 12 millions per year. A population of 345 million people in 1947 is expected to reach 525 million in 1970. This growth in numbers is **entirely** due to the falling death rate, which far outstrips the drop in the birthrate from 43 per thousand in 1947 to 37 per thousand in 1960. In other words, the terrifying increase in world population is due to the application of technology in the form of medicine, pesticides, agriculture, sanitation, water-supply and education. This has restrained those natural factors which have previously held the human population in check—excluding war. Far from being the result of an orgy of procreation on the part of the underdeveloped nations, the present population explosion is directly attributable to the application of technology, exported by the industrialised nations. The fertility of a nation is related to traditional attitudes and slow-moving social or economic factors, but the death rate has responded dramatically to technical advances. Major diseases such as malaria and bilharziasis have been brought rapidly under control since 1945. Improvement in water-supply, transport, diet and pest control have made considerable strides in recent decades. The famine predicted for India in 1967 did not in fact materialise because of a sharp rise in agricultural production and the distribution of food from abroad in time. But the net effect of these measures was to increase the total number of mouths to be fed, and to bring nearer the population crisis. Hundreds of millions of people—particularly in Asia and Africa—can expect to die of starvation before the end of the century unless better agricultural methods on a vast scale can be employed quickly, or the rate of increase of whole populations checked.

The effect of the growth of numbers is to be seen in many areas. First, there are far more elderly people in industrial societies. In 1951, the number of those over 85 in Britain was 224,000; in 1961 it was 338,000; by A.D. 2000 it will be 677,000. Thus the age-balance of society is altered, the 'problem' of old age becomes more widespread, and with a simultaneous increase in the number of young persons, the number dependent upon the middle age groups is much larger. The use of land for housing is another major consequence of the increase. It is expected that land for an extra 25 million people will have to be found in Britain by A.D. 2000—that is equivalent to 20 cities the size of Birmingham—some 6–7 million acres will be required for this. At the same time, food, water, mineral resources and fuel will be demanded at rates even higher than at present. Vast industrial and leisure development will be required, the present traffic problems and encroachment upon wild-life and unspoilt areas of country are bound to accelerate. Only the most careful planning on the part of those responsible for land and resources coupled with new standards of personal responsibility on the part of every citizen can hope to make the future bearable on these islands.

Of even greater awesomeness are the international problems which the growth of population will bring. Not only do millions already have to live under the threat of starvation, but strong expansionist pressures are likely to be developed by overcrowded communities. If these are coupled with the capability of using nuclear weapons, ideological warfare and racialist feeling, the threat to world peace may be great. China has set herself the target of a population of 1,000 million by the end of the century, but this could be catastrophic for herself, her neighbours and the world as a whole, unless the use of technology and racial attitudes are dramatically improved in a very short time.

Professor Dennis Gabor has argued, in his book **Inventing the Future**, that for a **prosperous** civilisation, even the Western population density is already beyond its maximum. Frustration and anxiety symptoms are widespread in our society. Over half the hospital beds are occupied by mental patients. It is little consolation to be able to escape from it all to a remote part of Wales, Scotland, France or Italy only to discover hordes of holidaymakers turning beauty spots into urban litter-heaps, deafening the sounds of sea and birds with city 'pop' on ill-tuned transistors, and polluting the air with petrol fumes. The planning of the future and in particular the control of man—his numbers, attitudes, relationships and standards—are the central tasks which the population explosion thrusts upon us.

2

The Control of Birth and Death

The population problem is perhaps the central issue facing the human race today. Compared with it, the problems of race, international peace, and space travel pale into subsidiary issues. We are threatened with gross overcrowding of the human species in the foreseeable future: a state of affairs which will lead to starvation, conflict and dehumanisation on a colossal scale.

It is generally agreed that overpopulation must be arrested by some form of birth control if disaster is not to overtake the human race. The limitation of the number of births by abstinence from sexual intercourse or by some artificial method is accepted by all responsible citizens.

In the open society generated by modern technology, with considerable emphasis on sex in advertising, mass entertainment and fashion, the path of abstinence advocated by the

Roman Catholic Church is heroic but somewhat unrealistic. Moreover, it stresses only the procreative aspects of sex. Particular attention has therefore been focused on other methods of birth control—some of which have been practised for many centuries.

There is evidence that the ancient Egyptians practised crude forms of contraception. It was certainly advocated by humanitarian reformers working in the slums of nineteenth century England long before pioneers like Dr Marie Stopes urged emancipated women to make use of new and improved techniques earlier in the twentieth. Now the intra-uterine coil or loop (I.U.D.) have been added to the rubber cap and latex sheath as relatively reliable mechanical methods. These are being widely distributed in India and Asia under Government schemes. Contraceptive chemicals have been available for some time, and since 1960, the contraceptive pill for women has been marketed as a sophisticated solution to the problem. The side effects of this newest method—which involves women maintaining an unnatural hormone balance for many months identical with that of the early stages of pregnancy—have sometimes proved unsatisfactory, and the long-term consequences including risks of thrombosis or liver damage have yet to be properly evaluated. But the 'pill' has contributed to the significant fall in the birth rate in Britain over the past five years.

Male sterilisation is practised on a fairly large scale in India among men with families to their credit, and a male contraceptive pill is likely to be marketed soon. The psychological problems of these methods are greater than others, since loss of virility—even temporarily—touches the male ego very deeply.

It cannot be denied that the use of any means of birth control introduces an artificial element into sexual intercourse, and in spite of the extent of contraceptive practise in industrial societies, aesthetic and psychological problems remain. Furthermore, no completely fool-proof method short of sterilization has yet been devised, nor—according to recent research—is it really desired.

In spite of much advice to the contrary from experts within his own Church, Pope Paul in his famous encyclical **Humanae**

Vitae (1968) has maintained on his own authority the traditional teaching of the Roman Catholic Church. Roman Catholics are urged to avoid all means of birth control except abstinence, *coitus interruptus*, and the use of the so-called 'safe' period; all notoriously unreliable procedures. The encyclical maintained that the true purpose of sexual intercourse is the procreation of children, and that the use of sex to express love and deepen the marriage bond is secondary to this. The pronouncement aroused dismay in many circles, and accented the problems of authority and conscience for Roman Catholics. It represented a somewhat inconsistent retreat from the modern world and its responsibilities in this matter, while advocating—rightly—the urgent need to improve methods of food production and distribution for the benefit of the hungry half of the world. Other Christian bodies have affirmed the legitimacy of artificial means of birth control within marriage if both partners find them acceptable in Christian conscience, thus placing the unitive role of human sexuality alongside the procreative role.

The availability of cheap means of birth control has increased the occurrence of pre-marital and extra-marital sexual relations, thus threatening traditional Christian standards of behaviour. No Church commends the practise of sexual intercourse outside marriage on the grounds that the act has the character of total, permanent commitment of two people to each other touching the deepest level of their being, and thus belongs properly to the married state. Moreover, the possibility of having children—even with the best contraceptive techniques—demands that sexual intercourse should be restricted to those who belong to a stable, permanent social unit. Contrary to much supposition, Christians hold a high view of sex, accepting fully its profound psychological and social implications. They hold that the application of technology in this intensely personal matter does not diminish these implications, even if present day attitudes lessen the social pressures on those who ignore them. Modern contraceptive methods therefore pose afresh the questions: What is the nature of human sexuality? What is the relation between the physical, personal and spiritual aspects of sex? What is the place of marriage in society? At the same time, Roman Catholics are required to attempt to reconcile the Pope's teaching about artificial means of birth control with the world population crisis, and thus face the problem of religious authority in a technological society.

Other means of controlling births inherited from antiquity are abortion and infanticide. The killing of newly born children was widely practised in Ancient Greece and Rome, but is particularly abhorrent today (as well as socially and psychologically disastrous) in communities built upon the rights and dignity of the individual person. Abortion—the removal of the growing foetus from the mother's womb during the early stages of pregnancy—is increasingly common. Eastern European countries, Hungary, and Russia, have particularly high abortion rates; so have Sweden and Japan. Legislation introduced in Britain in 1967 may have increased the numbers here. The case for this legislation rested on the burden to the mother and to society of the unwanted child, and the extent of illegal abortions. Particular stress is laid in the law on the careful evaluation of likely consequences of a birth on the health of the mother. With modern medical techniques, the termination of a pregnancy is a comparatively safe and simple operation from which the mother recovers—physically—quite rapidly. On the face of it, this appears to be a straightforward, humane solution to a major social problem. But the concern of those who disagree with the law—many of whom are Christian people—is for the psychological wound inflicted on the mother, and for the sanctity of human life, however embryonic. It is, they argue, difficult to decide in advance what the consequences of a live birth might be.

The number of abortions performed in countries where it is legal—statistics from other countries are notoriously unreliable—reveals serious inadequacies in the education of young people in personal and sexual relations. Hesitations in the United Kingdom about sex education and services offering advice in the matter of personal relations and birth control are based on traditional religious attitudes, but these attitudes have helped to place the burden of the abortion law and the unwanted child on British society.

The control of death now presents a formidable group of ethical problems. We have already noted how the fall in the death rate has been responsible for the rapid growth of the world population, and produced ageing industrial societies. Medical technology now makes it possible to prolong life long past the unaided limits by means of drugs, spare-part surgery, and artificial lung, kidney and heart machines. The removal of malignant growths, or decaying limbs, the suppression of lethal bacteria by modern antibiotics and the development of effective immunisation procedures against virus infections of many kinds have also made this far-reaching change possible. While there are biological limits to the length of human life set by natural ageing processes in body cells and organs, an expectation of life of 75–80 years is reasonable in industrial societies. Deformed and mentally subnormal people are now kept alive, and are often able to marry and have children, whereas a short time ago they would have died or been locked away and cruelly treated. But the fertility rate of the less well-educated is higher than that of more intelligent members of society. The fear has been expressed that unless this is changed by a widely effective birth control programme, or by improvements in the educational system together with a better physical and cultural environment for the poorer members of society, the average level of intelligence of the population will progressively deteriorate. F. C. S. Schiller expressed this fear when he wrote: 'humanity is destined to peter out in an overwhelming flood of feeble-mindedness'. If this threat is a real one, and not all educational psychologists would say that it is, the remedy lies in our own hands once again.

All these important changes—the ageing of the population, the increase in people dependent on the community medical and social services, the growth of numbers with greater educational needs—stem from modern medical technology. The enormous success of traditional medicine, that is to say the treatment of the sick, has given the community a new expectation of health. It has also drawn attention to the whole field of social and preventive medicine, until recently much neglected. Countries like Britain needed an industrial revolution and two major wars to take seriously the preservation of the nation's health. What sense is there in providing an expensive school dental service and permitting the sale of sweets on school premises? Why does health education and guidance on diet not form an essential ingredient in the curriculum of so many schools? Why has the incidence of venereal disease reached epidemic proportions in Europe, with over half a million cases of gonorrhea in 1963—in 1969 it was second only to measles on the Infectious Diseases list in Britain—yet at the same time there is a reluctance to provide adequate information about sexual relations or sexually transmitted diseases at home or in school? A major question facing advanced societies concerns the role of medicine in the community. What proportion of medical resources should now be devoted to preventive medicine?

New techniques raise further problems. It is the acknowledged duty of the medical profession to try to save life, but it can be argued that patients in many geriatric wards and nursing homes are being officiously kept alive. In connection with the problem of senility and with transplant operations: what is an adequate definition of death? Should doctors alone be responsible for decisions about medicated survival? Is not the high suicide rate among the elderly more sane than sinful? Should each citizen be prepared to offer organs from his body for transplant after death? Who should decide priorities for the use of limited numbers of new life-saving devices such as the artificial kidney machines? Should those with severe, inherited abnormalities or seriously subnormal intelligence be sterilised? In the face of the growth of population, and the need for a healthy society other questions also arise. Should the relation of doctor to patient be a strictly private, technical one, or should patients be treated in their social context? Is individual freedom to be asserted against the needs of future generations? Is it the job of the medical practitioner to mediate the requirements of society—as the priest used to do—or to attend simply to the felt needs of the patient?

These are but a few of the many questions which recent technical advance has thrown up. Some challenge traditional codes of conduct. Others demand a reappraisal of public expenditure and expectations from a medically self-conscious society. Moral choices are implicit in much medical treatment, and it is important that sentimentality should not blind people to the need for decisions in relation to the nature of human life, health or death. Christians, whose forefathers pioneered medical care in the Middle Ages and proclaimed the gospel of health through the missionary enterprise, have today a major responsibility of relating their conviction about the nature and total health of man in society—what the Bible calls 'salvation'—to the new situation of the overpopulated ageing world. Spiritual wholeness has to be related to individual physical and mental health **and** to the general condition of society; certainly the idea of the Kingdom of God in the teaching of Jesus is as large as this.

3

The Use of Resources

The rapid growth in the human population highlights the need for the resources of nature to be used as carefully as possible. First and foremost, the most vital resource of mankind is land. Land is the main source of food, minerals, space for living, working and recreation. The land supports the natural environment on which man unconsciously depends so much; from it he gets fresh water and re-oxygenated air. Although some have always fished or travelled the seas, the land has been the basis for man's greatest achievements and most necessary tasks. It is not surprising that he has always tended, like other animals, to be possessive of territory and committed to the earth.

Land covers nearly one quarter of the earth's surface. Of this only half is at present habitable, the remainder being either mountainous, desert or polar regions. Some 25 million square miles are available for living and cultivation, but of this 780 square miles are removed each day for roads, building, mining or by erosion. During the past century, the waste land on the earth's surface has increased from 10% to 25% as a result of man's intervention. With the growth of population, man dare not misuse the land. Although the most modern methods of agriculture have reduced the acreage need to support each individual to half an acre per annum, overall it is estimated that two and a half acres of good land per person are needed.

The most important element in the land is the thin layer on its surface, the soil. Soil is a highly complex thing, full of mineral and organic matter and huge numbers of living creatures. It has been estimated that the weight of creatures beneath the surface of a field is greater than the usual weight of cattle grazing on it! Soil science is a recent area of study, and comparatively little is yet known about it. But three major factors about soil are known and guide good land management: its liability to erosion; the sensitivity of its chemical balance; and its role in the basic water and nutrient cycles essential to organic life.

Erosion occurs constantly in nature, but it has often been accelerated by poor husbandry. The barren lands of North Africa, the Middle East and Southern China are the result of excessive cultivation, overgrazing, deforestation and unwise farming. The eroded areas of the central United States brought about by over-cultivation and wind erosion in the 1920s are reminders that careless application of new techniques can be as disastrous as the old ignorance. Neglect of hill pasture in countries like England and Wales, or over-grazing in hot countries like Africa are showing themsleves equally damaging for the soil and landscape today.

Chemical fertilisers and pesticides—now used on a huge scale to improve food production—have created new problems for soil conservation. It now looks as though the large-scale use of chemicals will not provide the final answers for food production. Experts have discovered that the application of chemicals creates massive biological problems in the soil in the long term. Better results are being obtained by careful land management using suitable crop rotation, natural fertilisers, and the judicious selection of suitable seeds for the soil and climate. Similar chemical and ecological problems have been created by the widespread use of chemical pesticides. These have found their way into the crops —and thus into the human system; into animals and insects essential for the proper ecological balance on the land; and into rivers and oceans. Certain of the persistent organochlorine pesticides, like D.D.T., D.D.D.—its derivative—and Aldrin, which accumulate in the fatty tissues of animals and birds, have a toxicity similar to that of asiprin; they represent a serious long-term danger to plants, animals and man.

Land supports vegetation and animal life. Trees, insects, birds and animals form part of a complex with air and water which has the marvellous capacity of renewing the soil to sustain the life-processes on which man also depends. Where that complex has been grossly disturbed—as by the deforestation of the Mediterranean lands over 3,000 years ago, or the removal of shrub and grassland in the Mid-West of the United States—the capacity of the land to support life has failed. Nevertheless, some changes can be wisely made—much of the present landscape of the Downs and grasslands in Britain is the result of man's activity. Conservation and restoration of forests is now increasing in the United Kingdom and in many parts of the world, though not rapidly enough to meet the insatiable demand for timber for building and paper. One Sunday edition of the **New York Times** uses timber from 150 acres of forest. The growing concern for conservation of land—grasslands, forests, wetlands, woodlands and high-grade farming land—from the encroachments of the mass of people reflects a quickening responsibility for tomorrow's resources. This sometimes creates conflict with city expansion, leisure interests and even the needs of the world's hungry.

Agricultural efficiency in the U.S.A. is now so high that farmers have been paid to conserve their soil rather than produce more food. The 1969 figures of the U.N. Food and Agriculture Organization show that food production in poorer countries rose over 3% faster than population growth. The exciting possibilities opened by the discovery of huge underground water-supplies beneath the Sahara Desert suggests that with intelligence and necessary capital some of the mistakes of the past could be rectified. It has been estimated that with this water, the Sahara could become the world's market garden, producing enough food to supply the present world population in a matter of ten years. Novel means of improving agriculture and using new foods from the sea such as those recommended by N. W. Pirie in his book **Food Resources, Conventional and Novel** are being explored in Japan and elsewhere. Thus the production of sufficient food and the wise use of land are not incompatible aims.

What is needed therefore to preserve the land to the many uses of man, is an ethical attitude towards it. As Aldo Leopold wrote in **A Sand County Almanac**:

> *A land ethic reflects the existence of an ecological conscience, and this in turn reflects a conviction of individual responsibility for the health of the land. Health is the capacity of the land for self-renewal. Conservation is our capacity to understand and preserve this capacity.*

The Christian conviction that 'the earth is the Lord's' reinforces this view of man's relation to the land. The certainty of the Israelites of the Deuteronomic period (*c.* 621 B.C.) was that the land in which they lived was God's, requiring their careful stewardship. Its produce was to be used for the benefit of all (Deuteronomy **14**. 28f; **20**.19; and **25**). These ideas were elaborated in the later Priestly Code compiled during the Jewish Exile (587–525 B.C.), requiring a sabbath year in every eight when the ground was to lie fallow and the inheritance remembered (Leviticus **25**.1–24). Christ's belief—expressed in the Temptation story

—that the earth could not rightly be possessed or ruled until its Maker's sovereignty was acknowledged, and that the fruits of the earth were for sharing, (Matthew **4**, **6**, and **14**) continued and deepened this profound tradition. It is expressed by the offering bread, wine and money in the Holy Communion service.

What is needed is that the truth of man's stewardship and responsibility to God and our neighbours—future as well as present —shall be translated into terms of practical use of the land, and into a positive attitude shared by city-dwellers as well as countrymen. This must challenge the policies of exploitation for quick economic gain pursued by so many. The waste and despoilation of the land are capital sins which will be visited on generations to come.

The same principle of conservation is required for other resources on the planet. The ruthless slaughter of the North American bison which took place towards the end of the last century, and the near extermination of most of the world's whale population in the 1950s are stern reminders of human folly and greed. The introduction of factory farming methods in the well-fed nations—to supplement the luxuries rather than the necessities of life—raises serious issues. The use of drugs to stimulate the growth rate of animals, and their confinement in dark pens in which they cannot turn round must surely cause concern on moral and medical grounds.

While the availability of coal, gas and oil appears to keep pace with man's huge and growing demand for energy, and these mineral resources are now being extensively supplemented with nuclear fuels, until recently there has been comparatively little concern about the pollution of the atmosphere and oceans by waste products. In the city of Pittsburgh, more than 7 million tons of coal dust are put into the atmosphere each year. Over Britain—a pioneer country in Clean Air legislation since 1863, when the first Alkali Inspectorate was set up—more than $1\frac{1}{2}$ million tons of smoke and 5 million tons of sulphur gases are released each year. The deadly smog which affected London and other British cities up to 1952 still hangs heavily over many industrial areas in the United States and Europe. Attempts are only now being made to deal with waste pits from British coal and iron mines of the last 200 years. The long term danger of petrol exhaust fumes and sulphur dioxide are only now being fully appreciated and some measures being taken to check them. The advance of metal refining methods, making possible the treatment of lean ores, and the development of new alloys or materials such as carbon fibres seems likely to meet the ever increasing demand for minerals in the industrialised states. But at the present rate, deposits of iron-ore are likely to be exhausted by A.D. 2500.

The supply of water fit for various human uses is now a matter of great national and international concern. Competing demands for land use, has, until fairly recently in Western countries, left behind the need for suitable water supply. Each individual in Britain uses domestically over 50 gallons of water each day, but industry uses more than half as much again per head. The quantities are expected to double by 1980. The controversies of recent years about water from Wales and the Lake District highlights the demand of the cities for fresh water. To this end the Water Resources Board and its 29 River Authorities in England and Wales are cooperating with some 500 water companies, the Meteorological Office, the Forestry Commission, the Water Research Association and other research bodies to develop a policy for water conservation. This involves the protection of land, lakes and rivers, and geological deposits; forming policies to control the provision of water and its uses for recreation as well as in agriculture, industry and the home. Industrial nations are now facing a major problem of water pollution in rivers, lakes and in the sea, as a result of the use of pesticides, detergents and fertilisers, and the dumping of waste products. In the U.S.A., the situation is recognised as grave, especially in the Great Lakes sites where much organic life has been destroyed. Pollution of the Baltic sea with D.D.T. is so bad that some species of fish caught there are dangerous to eat. The widespread use of detergents for washing and chemical fertilisers places a heavy strain on the rejuvenation processes of the soil through which polluted water passes. Perhaps the greatest problem confronting those anxious to maintain a clean and adequate environment on land and water is the ignorance and indifference of the public.

The author of the Book of Job marvelled at the stars, the beasts, the plants, the oceans, and at the skills and ingenuity of man. The author of Psalm 8 wrote:

> *What is man that thou art mindful of him,*
> *Or the son of man, that thou carest for him?*
> *Thou didst make him little less than God,*
> *Thou hast crowned him with glory and honour,*
> *Putting everything in subjection under his feet.* (R.S.V.)

Technology has, materially speaking, given man this dominion. He now has incredible skills and enormous power at his finger-tips; he can literally move mountains, see molecules and hear stars. But the question remains: what kind of dominion will man exercise? Will his rule continue to be one of short-sighted exploitation—as it has been so often in the past? Will it take account of man's responsibilities? Episodes like the Torrey Canyon affair, when a giant oil tanker ran aground near the Scilly Isles, spilling over 60,000 tons of crude oil into the sea near the English Coast, make this a questionable assumption. Will man's dominion be in the interests of the needy and less fortunate members of his society—let alone the rest of the human race? Will it be a rule of waste and self-indulgence at the expense of future generations? Will it be the dominion of Christ or Mammon? Ultimately, today's technology thrusts that question upon us. Western man, nurtured in Christendom, from which the dynamic of modern technological society has come, must answer that question for the sake of tomorrow's world.

4
The Challenge of Tomorrow's World

Three major issues are presented by modern technology to our society. They are the need for sound planning; the need for new standards of personal responsibility; and the need for new aesthetic standards to be applied to man's living and environment.

The only possible way in which the world's population can be fed, employed, educated and allowed to develop through work, relationships and play, is by wise planning. Planning aims to use properly and to maintain the supply of resources needed for life in a balanced, healthy environment. Planning uses the administrative skills developed over the centuries, and particularly in industry and government at every level, but it is more than mere bureaucracy. It is the creation of the most favourable interaction and balance of factors needed for the greatest quality of life in the future: it is the politics of tomorrow. All modern industrial societies are, to a greater or lesser extent, societies committed to planning—if not yet planned societies.

In most cases planning is a major concern of national government, and it is to be hoped that it may become the concern of regional and local government—indeed of the whole population—more and more. Nowadays, planning goes far beyond the early notion of town and country planning embodied in legislation which began in Britain in 1909. The Acts of 1943 and 1947 laid down the main aims of planning of this type: (1) to shape planning policies for each area; (2) to bring development under the control of planning authorities or central government; (3) to give powers and financial aid to local authorities for planning purposes; and (4) to provide for special amenities such as historic buildings, trees, play areas, and nature reserves. In Britain there now exists a highly complicated development planning system—as anyone who has had to erect a new house or garden shed knows. Under the former Department of Economic Affairs, Regional Planning groups have been set up and have issued major reports on their respective areas. But alongside this, there is a complex system of conservation organisations, ranging from the Natural Environment Research Council to local Naturalist Trusts. There are agencies concerned with water, power, farming, transport, minerals from the level of government departments to urban or county interests. All are concerned with planning, though some in the restrictive sense only.

It has been pointed out by Robert Arvill in his important book **Man and Environment**, that major weaknesses in Britain are a shortage of information and of planners—personnel suitably qualified to obtain and evaluate information and to project appropriate schemes on the strength of it. Modern planning is a highly skilled task involving much imagination, and a familiarity with many disciplines—population statistics, economics, geography, architecture, agriculture, traffic engineering, sociology and salesmanship. At the same time, planners in a democratic society have to be able to work with—and under—laymen in public authorities, and in full view of a critical public. Happily, public discussion of planning problems is now developing in Britain and other countries, although far too much planning and effective decision-making still occurs behind the scenes of central and local government. Plans for regional government in Britain seem likely to mature in the 1970s, and an upsurge of concern for democratic participation among young people may find an outlet here.

To many people, planning is regarded as the antithesis of freedom. Political propaganda has for some years implied that the responsible public control of resources of land, water or amenities constitutes an infringement of individual rights and opportunities. For those with ample means, property or investments, planning certainly may be restrictive of private ambition, but at the same time it makes available to all, including those interests, greater choices in the long run, and greater responsibilities and freedom. Planning cannot of itself create Utopia, nor can the concept of planning suggested above be equated with any political ideology. Planning is the collective ordering of community affairs for the common good, and for the sake of posterity. Only through intelligent planning can the individual make ends meet and enjoy life to the full, and only **community** planning will make possible—for those who wish to avail themselves of it—the best life possible for the community as a whole in a responsible environment. If we are concerned about the quality of tomorrow's world, and about the greatest opportunity for a full life for the greatest number of people in this country and throughout the world, then planning there must be.

Christians, like many others, tend to be hesitant about planning the future, even when they order their own affairs with prudence. To them, planning smacks of impudence; taking God's place in the ordering of history; or presuming to shape the future by merely human insights. It runs the risk of identifying human Utopias with the Kingdom of God. But the planning demanded by modern technological society is not designed to give complete control of the future; in any case far too many uncertainties remain for that to be conceivable. Nor does it remove from the individual member of a planned society responsibility, choice or human dignity. Compared with previous, unplanned development, democratic planning heightens collective responsibility and gives real choice to more people, now and in the future. The history of the industrial revolution has shown that ignorant waste and spoilage reduces tomorrow's possibilities. Further, if man is meant in some sense to be like God—as he is called to be through Jesus Christ—then to participate in wise planning of resources and opportunities is to become Christlike and godly. The form of Christian living today and tomorrow must be shaped by the responsibilities given to mankind, of which planning is undoubtedly one. Christ's love of the world therefore involves Christians in this positive attitude towards God's creation: in real stewardship of the future. It is high time the churches recognised this, and acted upon it.

Modern technology clearly demands new standards of personal responsibility. In President Johnson's Message to Congress in

February 1966, he set out a 'creed' to govern man's attitude towards the natural environment. In this he included:

> *the right to clean water—and the duty not to pollute it; the right to clean air—and the duty not to befoul it; the right to surroundings free from man-made ugliness—and the duty not to blight it; the right to access to the beauty of the countryside—and the duty to preserve it; the right to appreciate plants and animals in their natural habitats—and the duty not to exterminate them.*

Codes of this kind represent a first step in areas of responsibility which modern technology has opened up. But more is required. It is necessary for people to weigh the implications of new technical developments—good and bad—on other people and on the environment. Too much recent technological advance has occurred without due attention being given to appropriate disposal of waste products for example—or the long-term consumption of minerals. The throw-away attitude of the 'Consumer Society' has been encouraged by skilful salesmanship on behalf of commercial interests, but it has met with little protest from a bemused public. Poor standards of production, design and service are commonly accepted. In capitalist countries, the profit motive, the size of the wage packet and the national balance of payments have been allowed to dominate personal and political interests. Yet the long-term policy, if it is to safeguard the well-being of present and future generations, must strike a balance between profit and reliability, instant affluence and conservation, economic success and international responsibility. The emergence of the mass media requires higher standards of discrimination throughout the population—or leisure will become synonymous with cheap entertainment. The arrival of the motor car places a lethal weapon in the hands of millions of people, demanding increasing levels of skill and responsibility by drivers. These are all matters of ethical and religious concern today.

These issues focus upon the kind of education needed in a

technological society. They make it clear that what is essential is not only an education into the disciplines, skills and techniques of an international, technical and highly competitive world coupled with the personal development of which each is capable for the full enjoyment of life and leisure, but also an education into personal, social and political responsibility, informed by imagination and illuminated by strong conviction, care and hope. Christians will recognise here St Paul's trilogy of faith, hope and love: they would argue that only on the basis of such commitment can we welcome the new dignity open to man in the new era of technology.

Thirdly, the present situation calls for new aesthetic standards. The quality of art, design, architecture and music of the last 200 years generally leaves much to be desired. The squalid factories, scruffy stores and jerry-built houses of many industrial towns present a dismal prospect. The sordid suburban sprawl outside many towns and cities in Britain and America reflects a terrifying indifference to beauty. Far too many products of today's technology continue this tradition. Cheap plastic knick-knacks, flowers or kitchen gadgets jostle with poorly designed clothes in overcrowded stores. Those anxious to uphold well-tried standards of art, music and design are frequently regarded as archaic old-timers. But new materials and a new cultural situation create new possibilities for aesthetic exploration, and the signs are that improvements are taking place, especially in commercial art and architecture, although there are no grounds for complacency.

This is more than a private, subjective matter. Beauty cannot rest wholly in the eye of the beholder, since it has qualities of integrity, appropriateness and form which can be appreciated by all who stop and reflect with sympathy. Art and technology—once married in the age of the craftsmen—can and must be reunited if the growing man-made environment is not to degrade those who live in it. Beautiful things exalt people, show them their potentiality, and educate their imagination; and imagination is a faculty more needed in an age of technology than ever before if man is to survive.

In the Christian tradition, beauty has been accorded a high place since St Paul's exhortation to think on such things. This echoed a noble element in Jewish as well as Gentile culture in which beauty was associated with holiness, goodness and truth. The Church has often found itself cast in the role of patron of the arts, and preserver of standards in architecture, music, sculpture, painting and drama. During the industrial revolution it gave in too readily to 'worshipping the Lord in the beauty of ugliness', and aesthetic standards among Christians are often poor still, as their churches and services testify.

If tomorrow's world is to be a beautiful place, effort, expense and education will be needed. More important, deep convictions will be required about the value and meaning of beautiful things. What have the different religions to offer for the encouragement of standards of beauty? Are there any criteria upon which intelligent people of any religious belief or of none can agree for aesthetic standards? What more should be done by government at every level to ensure better standards in the style of the artificial environment and social life?

5

Topics, Projects, Questions and Sources

1

Prepare an exhibition to show the growth of world population by continents (or major countries) since 1920. Indicate, where possible, changes in birth and death rates, and point out the major factors responsible for these changes.

2

Make a comparative study of the change in land use in an area covered by a radius of 5 miles from the centre of a medium-sized town. Indicate land used for agriculture, woods, commons, houses, parks, industry, roads and railways and base your comparison on information for (say) 1850, 1900 and 1950. Find out from the local planning authority what the situation is likely to be in A.D. 2000.

3
Outline and discuss the teaching on birth control in the papal encyclical **Humanae Vitae**.

4
Has anyone the right to deprive an unborn child of life? Discuss.

5
Find out the proportion of people over 60 years of age in the population of your town. What special provisions for old people are made by *a*) national government and *b*) the local community? What contribution can older people make to the local community?

6
Arrange a discussion with a doctor **or** visit a local health clinic to discover what you can about one or more of the following:

a) local provision for preventive medicine,
b) problems of sexually transmitted diseases,
c) psychiatric services in the community.

7
Arrange a group display to show the policies and inter-relationships of organisations concerned with planning and conservation in your area.

8
Trace the idea of stewardship through the Bible.

9
Read and discuss **1984** by George Orwell or **The Question Mark** by Hugh Montefiore.

10

Make a list of qualities you would consider appropriate to the notion of a beautiful building, piece of furniture, item of clothing or picture. Compare and discuss your list with that of another person. In what ways can aesthetic standards be raised in our society?

Sources:

General—

(Ed) E. Barker **The Responsible Church** *S.P.C.K.* 1966.
R. Arvill **Man and Environment** *Penguin* 1967.
D. Gabor **Inventing the Future** *Penguin* 1963.
J. A. Lauwerys **Man's Impact on Nature** *Aldus Books* 1969.

Population—

The Report of World Population Prospects 1963 *U.N.E.S.C.O.* 1964.
Registrar-General's Review of Population Statistics, 1966 *H.M.S.O.*

Control of Birth and Death—

G. R. Taylor **The Biological Time Book** *Thames and Hudson* 1968.
D. Gabor **Inventing the Future** (Ch. 6) *Penguin* 1963.
Colin Clark **World Population** In *Nature*, 181, 3 May 1958.
Humanae Vitae Papal Encyclical 1968 *Catholic Truth Society.*
Pacem in Terris Papal Encyclical 1964 *Catholic Truth Society.*

Use of Resources—

R. Arvill **Man and Environment** *Penguin* 1967.
N. W. Pirie **Food Resources Conventional and Novel** *Penguin* 1969.
Technology and Economic Development
A Scientific American Book, *Penguin* 1963.
A. R. Ubblehode **Man and Energy** *Penguin* 1963.

Planning Responsibility and Aesthetics—
The Buchanan Report **Traffic in Towns** *H.M.S.O* 1963.
Regional Development Plans 1966 onwards *H.M.S.O.*
R. Arvill **Man and Environment** (Chapters 16–22) *Penguin* 1967.
L. Mumford **Art and Technics** *Columbia Press* 1952.
D. Jenkins **The Educated Society** *Faber* 1966.
D. Whittle **Christianity and the Arts** *Mowbray* 1964.
H. Montefiore **The Question Mark** *Collins* 1969.

C
The Impact of Technology on Religion

In the first part of this book, it has been argued that religious convictions have in the past helped to shape human technology from the manufacture of hunting tools to the construction of the city. It has been suggested that the dynamic movements of expansion, innovation and industry of the sixteenth and seventeenth centuries—the precursors of the industrial revolution proper—were inspired by the rejuvenated convictions of the Reformation and Counter-Reformation alike. The justification of the discussion in the second part of the book rests on the assumption that it is possible for mankind to control the application and development of technology by moral and religious criteria.

It has been argued that in the course of human history, man's technology has shaped his culture and affected the form, if not the content, of his religious beliefs and practices. There can be little doubt that this process is taking place today in Christianity and other religions. The term 'secularisation' has been coined for the subtle processes whereby the springs of traditional beliefs and practices have been eased in advancing technological societies. The phenomenon has been detected in the Old Testament by scholars like Professor Gerhard von Rad as well as by contemporary observers of our society such as Professor Harvey Cox.

The term secularisation is a difficult one and is used in many contexts. The actual extent to which agnosticism and atheism have replaced orthodox Christian belief is hard to assess, and some sociologists like Dr David Martin have recently criticised some of the assumptions and conclusions of the 'secularisation' school. But we are faced with a complicated situation, full of cross-currents and new pressures, some of which undermine religious concerns, and others which may reinforce them. Personal religion, Christian unity and fellowship appear to have been strengthened in response to—and sometimes as a direct result of—recent technical and social changes, whereas attendance at formal worship, observance of certain moral standards and respect for some hallowed authorities, institutions and traditions seem to have been weakened.

In spite of the obvious complexity of the situation, and our present lack of clear information about such intangible matters as religious belief, personal attitudes and moral insight, among the population as a whole in the past as well as the present, it is evident that the age of modern technology is having a considerable impact on Christian thought and behaviour. The Christian Churches without exception are having to review the meaning, emphasis and consequences of their traditional beliefs, and reconsider patterns of acceptable behaviour. A new concern for religious education and for the relevance of faith and worship to the modern world has emerged. In the short discussions which follow, an attempt is made to indicate some of the problems which technology poses for Christian belief and practice today.

I

Reshaping Christian Faith

In the Book of Common Prayer of the Church of England, special prayers are provided for occasions of drought. The belief of many, inherited from earlier centuries is that the Lord of the heavens will, in his providence, answer such prayers and send the necessary rain. Until recently, man had no control whatever—so he thought—over the climate. Now it is recognised that man's interference with the landscape has considerable consequences for the water cycle; the vapour trails of high flying aircraft may well affect the cloud pattern; experiments with certain iodides dropped from aircraft over hot regions suggest that he can stimulate directly the formation of rainclouds. Israeli irrigation engineering in the Sinai Desert and British efforts in the Sahara have begun to bring to fulfilment the ancient hope that the desert should blossom like the rose. But modern technology offers more than this. U. Thant, Secretary General of the United Nations said this in 1962, in a speech connected with the launching of the Development Decade:

> *The truth, the central stupendous truth, about developed countries today is that they can have—in anything but the shortest run—the kind and scale of resources they decide to have. . . . It is no longer the resources that limit decisions. It is the decision that makes the resources. This is the fundamental revolutionary change—perhaps the most revolutionary mankind has ever known.*

In the light of what has been said about the limits of certain of the world's resources, this may seem an optimistic statement. But the synthesis of new materials and food, and the production of energy from nuclear fission, point to the truth of U. Thant's statement where the organisation and know-how exist for such decisions to be made. If it is not yet fully true, it certainly will be in Europe and in the U.S.A. within a decade. And the implications for man's existence are stupendous.

In a perceptive essay in the symposium **Spirituality for Today**, John Townroe has made a careful evaluation of the impact of technology on the notions of prayer and providence. He illustrates his argument by comparing the exhortation to keep Days of Humiliation by the absent vicar of Warminster to his parishioners on the outbreak of Cattle Plague in 1857, with the methods used to deal with a major outbreak of Foot and Mouth disease today. Man has gained control of many areas which previously were assigned exclusively to the Providence of God. This has shaken the belief of many people that the God with whom we have to do is 'real'. Moreover it appears that he cannot often be persuaded to intervene directly in the natural order to alleviate suffering or disaster. This therefore places a large question mark over the nature of penitence and prayer—especially the prayer of intercession.

The re-examination of the New Testament in the light of this dramatic change in man's relation to the environment has brought certain features of the Gospels and Epistles to light. Notably, that the kind of care, protection and blessedness on earth which was the longing and expectation of many Jews, and is the desire of many Christians still, fails to take account of the kind of providence shown by the God and Father of Jesus. Jesus lived a life of blessedness not as successful, prosperous, carefree individual, but in poverty and compassion for others. Faced with agony and death, he prayed '. . . nevertheless, thy will be done'. Salvation is seen to belong ultimately to the realm of resurrection, and the Gospels show the suffering and death of Jesus Christ as providence incarnate. Many false ideas of God's relation to man are challenged by this—as they are by the new situation in which man finds himself. God is a God who transcends his creation, he is not to be conceived as a God who is part of creation, a being or force within the system whose control over the universe is like our control over things.

Similarly, the notion of prayer has had to be reconsidered. The possibility of God's intervention in expected or unexpected ways is seen by some to follow logically from the notion of God as a loving, purposeful God on whom, as the ultimate reality, all existence rests. Others, such as the philosopher D. Z. Phillips in his book **The Concept of Prayer**, argue that we cannot sustain the view that God intervenes in the world at our request without placing God among the objects of existence and making him unworthy of belief. Intercession has therefore to be understood as an expression of trust and concern to God, rather than an exercise in twisting the Almighty's arm.

Man has always wanted to shape his own destiny and master his future. The uncertainty of tomorrow has led him to try at least to know what it will be like. Technology has given him a glimpse of control over the future—as yet only slight, but enough to stimulate his desire afresh. Rational, technological man is paradoxically very superstitious. The more organised he becomes, the more he gambles and speculates on the future. In particular, horoscopes and bogus religious superstitions about the future flourish. Industrial society has thrown up a remarkable range of semi-Christian sects offering bizarre expectation for the future, and attracting many who have little understanding of the Christian hope. The New Testament insights into the role of man as a fellow-worker with God in his coming Kingdom; of God as the sovereign lord of time and creation, yesterday, today and tomorrow; together with the thrust of the Christian understanding of history towards a meaningful goal, provide a wholesome basis for prayer, hope, and energetic technological activity alike in this situation. From such insights a coherent frame of reference for the responsible use of man's technology for the service of humanity and for truly spiritual ends can be derived. But it has to be admitted that the churches have so far only just begun to tackle the task of working out a theology of hope, and a workable ethic of technology.

The attitude towards the natural world generated by technology requires Christians to review their doctrine of creation. Concern for the natural world, and the relation of God—and therefore man—to nature, have often been lost to urban Christians. Since the rise of modern science, and the apparent removal of the Almighty to a place beyond the model of the universe opened up by astronomy and mapped by the supposedly final laws 'governing' its behaviour, the emphasis in western Christianity has shifted to the realm of feeling, of personal salvation and forgiveness. A glance at the hymn books shows this trend. The search for personal significance in developing industrial society, coupled with the close association of religion with the social standards of the middle classes tended to make Christianity into a religion concerned with guilt, inner consolation and social conformity. The heroic efforts of western missionaries has often been limited by their inability to dissociate the Gospel from this background. As a result, Christian concern for creation and its right use by man has not figured prominently in the proclamation of the Gospel of the western churches. Much has occurred in exploitation of man and nature without protest or constructive criticism from the Christian community. The indifference of Christians to the created order has been shown by the ugliness of many of their buildings, the insensitivity of many of their services.

Now, however, a new awareness is appearing, and a new concern for God as Creator and his purpose in creation is evident. From this concern for the spiritual meaning of the natural world, there are signs in the pronouncements of the Vatican and the World Council of Churches of a serious attempt to grapple with the problems facing man in the created order.

Among many exhortations, the 1968 Assembly of the World Council of Churches said this:

> *Technology is a radically new way of exercising dominion over the earth. It is a discontinuity in our million-years of history. So it is vital for theological thought to address itself to the constructive and destructive potentialities of this central source of change, as well as its visible consequences.*
> *Theological thought can only meet this challenge if those in administration, industry and technology join forces with the theologians in working out the response. They must give knowledge and receive vision.*

Sections 45 & 47, **Section on World Economic and Social Development**

And in the Message of the Assembly:

> *The biblical message is that man is God's trustee for creation, that in Christ the 'new man' appears and demands decision.*
> *Therefore, with our fellow men we accept our trusteeship over creation, guarding, developing and sharing its resources.*

Faced with the hungry millions of the world, and the economic injustices of today, the telling prayer '. . . give us our daily bread' is seen to imply a new relationship of man to nature and his neighbour, and a new understanding of what the Lord's prayer calls 'our trespasses'. In short, technology forces Christians to take fresh account of the God they profess to believe in as Creator, Redeemer and Lord, and of the behaviour and attitudes he now requires.

2

Remoulding Christian Practice

The advent of the city and the motor car have dramatically altered the pattern and concerns of local church life. In place of the parish ideal of a single centre of worship set in the midst of a community and related to its work and play, the city situation has separated home, work and leisure interests. The motor car has broken the limits of a sabbath stroll; and the traditional hours of worship have shown themselves unsuited to the social habits of today. Many city churches gather congregations from a wide area, often far outside the parish boundaries. Churches in the inner zones and industrial areas have fallen on hard times, while suburban places of worship—usually stressing the domestic implications of Christian life and offering fellowships for young people, women and the elderly—tend to flourish. Gibson Winter, a North American commentator, has referred somewhat scathingly to 'the suburban captivity of the churches'. In some places, notably the United States, Sunday evening worship has largely disappeared as families have taken the opportunity to travel to the country after morning church and spend the rest of Sunday away from home together. A similar trend can be observed in Britain, especially in the South East.

The development of modern technology has indirectly affected the role of the minister or priest. Whereas in earlier times he was related to the whole life of his people, now he meets them more often than not in the domestic and leisure situation. While he used to be guide, mentor and friend for every circumstance, now he often finds that others with new expertise in medicine, psychiatry, social welfare, probation work and family case-work have arrived before him. His social role is essentially that of the wise amateur, the general practitioner of concern and care. But this difficult task is not always understood—even by those in the ministry. Serious lack of definition of the work of the local church ministry has led to a sharp decline in the number coming forward to serve in it. On the other hand, new 'specialist' ministries have emerged in recent years and there seems to be little shortage of men anxious to serve in the apparently more precise role of school, hospital, college, prison or industrial chaplain. The Church has been obliged to take the new groupings of urban society seriously, and to seek to influence the new centres of power of the changing social order: the mass media—especially radio and television; industrial management and the trades unions; the medical profession; the education complex; and central and local government itself. In the new cities, prominence has been given to central churches and cathedrals, and in places like Coventry Cathedral significant new patterns of ministry and influence have been set up.

It is in the sphere of moral standards and social attitudes that the open situation of modern, technological society have challenged the Churches most directly. The issues of birth-control and extra-marital sexual relations has been discussed above (Section B). Problems of standards on television, care for the elderly and unwanted children, the moral issues surrounding modern warfare and the nuclear deterrent, attitudes towards people of other races and cultures all require consideration and resolution if the Christian claims are to appear credible and effective today. At the same time, Christians are challenged to participate responsibly in their society—a difficult and dangerous task in some countries in the world, and a demanding one even in the democratic West. Some of the traditional views of vice and virtue hardly constitute an adequate yardstick by which moral responsibility can be measured in today's society. Personal application of the Ten Commandments—beloved of the average Englishman as a substitute for Christianity—do not provide the kind of social code which the love of Christ implies today, important as some of those ancient precepts may be in context. The skilful balance of guidance, knowledge, and freedom to take responsibility in this day and age requires insight, understanding, care and leadership. The Churches are finding opportunity at the international level to give these things to those who will hear. The sensitivity of political leaders to the pronouncements of the Vatican and the World Council of Churches shows this clearly. There are some signs that national and local churches are trying to extend this lead for the moral and personal issues of our time facing ordinary folk.

3
Reviewing Human Nature

The incredible picture on television, seen by millions of viewers throughout the world, of two men setting foot on the surface of the moon shows the new humanity to which Christians have to speak. Technological man differs from his predecessors in certain key respects. Unlike pre-industrial man, he is not content with his knowledge, his powers or his station—he is not submissive to authoritarian structures. Unlike early industrial man, he is not preoccupied with work, with his past or with his sinfulness and feelings of guilt. Unlike his nineteenth-century predecessors, he does not subscribe to final solutions, dogmatic answers or imperial pretensions. He is a critical expert ready to cooperate with others, interested in himself, his world, his future; a man determined to enjoy experience if at all possible, even if his society and situation make him subject at times to great loneliness, uncertainty and depression. Christians need to take stock of technological man if they are to relate their faith and practices to his condition.

Modern technological man seems almost to be a new species. He is capable of living in outer space or in the depths of the ocean; surviving for years in Antarctica, or living a cool, sophisticated life in a desert city. His powers seem vastly different from those of prescientific man. He can calculate huge sums in a moment with computers, speak over thousands of miles, see into the world of the cell and hear the strange music of the stars. Yet he remains basically an ordinary man, like his forefathers. He has the same psychological and emotional needs for love and attention, the same difficulties in making decisions, the same feelings of shame or pride, the same beginning and the same certain end. In terms of his essential being, therefore, technological man remains just a man; a weak, fallible, dependent creature of brief life-span. His personality, needs, fears and conscience are the same fundamentally as before. He still reaches out to other persons in his relationships, and to the strange dynamic, spiritual mystery which surrounds and encounters him in the universe which he often calls God. He still attempts to do this through gesture and language, art and symbol, thought and science, religion and morality, and through technology. Man himself is still a great mystery: the mystery of his life with others in a mysterious universe remains his deepest concern.

For this reason alone, if for no other, man's ideals, hopes and goals are properly limited by a concern for persons set in a universe which he can only just begin to understand. Respect for human personality—and that means other people's as well as his own—and respect for what he knows of reality are the only basis for human ideals or actions which can be pursued with integrity. The yardstick of human freedom, dignity and potentiality, combined with truthfulness about himself, his fellows and his world, has to be applied to every dream, and every use of technology to achieve it. By this measure, religious dogmas, political ideologies, social codes, economic systems, scientific maps and psychological patterns must be judged: so must social organisations, ways of work, media of communication and other modern technologies. But by themselves, none of these things can provide a final or adequate criterion by which to assess man's dreams or deeds; only the mystery of personal life in a mysterious universe with a spiritual as well as a material dimension can do that.

The justification of this approach has been given to Christians in the life and work of Jesus of Nazareth; in what led up to it in the Old Testament, and in what followed in the New Testament and the life of the Church. The figure of Jesus, as the man of hope, power and compassion, is for Christians the prototype of modern technological man. He gives the essential clue which this perspective seems to require in a society which looks for a personal model and a meaningful goal. But whatever justification is accepted by Christian, Jew, Mohammedan, Buddhist or man of any other faith, the categories of persons in relationship, of creatureliness, of integrity and of respect for the many-sided nature of reality must surely govern man's science and technology. These categories must make up the model for life in the era of technology or else, like Faust, man will have sold his soul.

4
Further Questions and Sources

1
Does technology make religion superfluous?

2
How would you expound the doctrine of creation?

3
'Prayer is not an exercise in twisting the Almighty's arm, but an expression of care and trust'. Discuss.

4
What do you understand by the term 'the supernatural God'?

5
How should the Church use her influence in modern society?

6
Is technological man a new species with important new characteristics? If so should the Church tailor its message to his need and nature?

Sources:

D. C. Munby **The Idea of a Secular Society** *O.U.P.* 1963.
H. Cox **The Secular City** *S.C.M.* 1965.
D. Martin **A Sociology of English Religion** *S.C.M.* 1968.
D. Z. Phillips **The Concept of Prayer** *Routledge* 1967.
P. Baelz **Prayer and Providence** *S.C.M.* 1967.
D. Jenkins **Guide to the Debate about God** *Lutterworth* 1967.
G. Winter **The Suburban Captivity of the Churches** *S.C.M.* 1965.
K. Slack **The British Churches Today** *S.C.M.* 1969.
M. Keeling **Morals in Free Society** *S.C.M.* 1967.
P. Tillich **Theology of Culture** *O.U.P.* 1964.
R. Shinn **Man: The New Humanism** *Lutterworth* 1968.
A. McIntyre **Secularization and Moral Change** *O.U.P.* 1967.

Index